THE BULLWHIP'S ALGORITHM

THE BULLWHIP'S ALGORITHM

ARTIFICIAL INTELLIGENCE AND AMERICA'S RACIAL CASTE SYSTEM

KAMAU BOBB

BEACON PRESS, BOSTON

BEACON PRESS
24 Farnsworth Street
Boston, Massachusetts
www.beacon.org

Beacon Press books
are published under the auspices of
the Unitarian Universalist Association of Congregations.

Printed in the United States of America

29 28 27 26 8 7 6 5 4 3 2 1

This book is printed on acid-free paper that meets the uncoated paper
ANSI/NISO specifications for permanence as revised in 1992.

Text design and composition by Kim Arney

*Library of Congress Cataloguing-in-Publication
Data is available for this title.*
Hardcover ISBN: 978-0-8070-2343-3
E-book ISBN: 978-0-8070-2342-6
Audiobook: 978-0-8070-2570-3

The authorized representative in the EU for product safety and
compliance is Easy Access System Europe 16879218, Mustamäe tee 50,
10621 Tallinn, Estonia: https://beacon.org/eu-contact.

For my dad, Robert Arlington Bobb,
the origin of the Cartesian plane.
I have emancipated my mind,
and kept the faith. Peace.

CONTENTS

PROLOGUE

YESTERDAY AND TOMORROW

The Sankofa bird is a Ghanaian symbol that represents the importance of looking into history and seizing the lessons there as preparation for the future. The message is timeless. It is always important for a people to know their past, to be anchored to a common understanding of who they are and how they came to be. That ability to look back and know is necessary for a people to evolve with purpose, rather than be subject to the whims of the wind, false prophets, and the randomness of tomorrow. For Black people in the United States, the Sankofa bird is a kind of protective amulet. Our history is fraught with tragedy, terror, and triumph. To not know and not heed the lessons from our American story is dangerous. In this season of artificial intelligence (AI) in education and the evangelism it's wrapped in, the Sankofa bird helps us to see more clearly. Big Tech and edtech companies are making grand promises about what AI will do for education—claiming it will catalyze change, improve teaching and learning, increase effectiveness and efficiency, promote equity, offer personalization, close gaps, right wrongs, and redefine the institution of school. They say, in emphatic colorblind terms, that it will be good for all. Determining whether any of these promises are real or merely bait for the desperate to remain damned requires clarity of sight. It calls for foresight sharpened by hindsight. The ability to not get swept away in the evangelism of AI in education requires looking into the past. It requires being tethered to lessons learned about what wrongs need to be made right, what gaps need to be closed, and what guilty party created them. Black people have been dependent on and abused by American education for generations. Now, not only is the content of education under withering assault through comprehensive attempts to erase the Black story from the American story, but the promises of AI technology as a cure-all are bearing down with relentless intensity. The risk is of being lured into accepting that AI in education can solve problems that technology did not create. If we're not careful, it can serve as a decoy to let the guilty party off the hook and remain a repeat offender. Without the lessons of history to light the path ahead, we can be seduced by the awe of AI as it turns our attention away from the American race demons that are, in fact, the guilty party.

History is not far away. It's close. The year 2026 marks the 250th anniversary of the founding of the United States. It is also the year that I turn fifty-four years old. I am not an old man—I like to say I am in the youth of my old age—yet I am of the first generation of fully free Black people in American history. Slavery began in 1619 and predates the founding of the United States. It was a continuous American practice until the signing of the Emancipation Proclamation in 1863. It didn't end in fact until the Civil War was over in 1865. The post-war freedom of American Reconstruction was short-lived, from 1865 to 1877. Save for the children born in those dozen years, of the approximately four million Black people in America at the time, nearly all were born as slaves. They were born as other people's property, and their freedom was theoretical. They had virtually nothing to safeguard it. Reconstruction was followed by nearly a century of Jim Crow–era racial apartheid—a legal regime that enforced the second-class citizenship of Black people in America. That system came to an end after the blood-soaked civil rights battles culminated in the Civil Rights Act of 1964. That was, finally, the end of legal discrimination against Black people in America. Two hundred forty-six years of slavery followed by nearly a century of Jim Crow apartheid—1619 to 1964. Three and a half centuries of explicit legal oppression and racial terror ended in 1964. I was born Black in 1972, eight years later, in Brooklyn, New York.

That sense of freedom is neither just the abstract stuff of history and chronology, nor is it limited to the United States. I am the first person in my family to be born in America. My family is from Guyana, in South America. Guyana, like all of the African diaspora, was a slave and colonial state since the early seventeenth century. Various Europeans traded spurs over our dominion across the centuries, with slavery being the common throughline. The British were merely the last in line. Slavery in the British Empire ended with the Slavery Abolition Act of 1833, but Guyana remained a colony for more than another hundred years until it gained independence in 1966. My father would say he is an African subject of the British Crown born in a strange land. In his mind, the idea of genuflecting to the Crown, of bending in front of the Queen of England, was a marrow-deep insult to the memory of who we are. It was a living, ever-present component of my development. That sense of firstness—first to be free, first to stand up straight, first to not be subject—closed the gap between memory and identity in my upbringing. As such, memory is the bedrock of my identity. It leaves me deeply appreciative and respectful of national milestones and how they serve as markers of nations' memories of themselves. And for Black people, the American story and the diasporic story are interconnected. The Civil Rights Act of 1964 was a mere two years before Guyanese independence in 1966. We were emerging as free people for the first time, together.

The 250th anniversary of the founding of the United States is the kind of milestone that forces the nation to look back and ahead—to consider yesterday and tomorrow. In the current environment where America is trying to make itself great again, there is the obvious question: *Great for whom?* Sadly, this anniversary pits discernment, critique, and learning past lessons against the frothing full-throated mantra of America First—the idea that the nation has only ever been great, regardless of the generations of people that have suffered enslavement, slaughter, and subjugation in its construction. Given that I am of the first free-born generation of Black people, from my point of view, there has never been a time when America was fully great. We've only ever been on the steady march to achieve a more perfect union. Alongside criminal justice and so many other areas of American life, this is especially true in education for Black people. American educational history is overflowing with abuse, segregation, substandard structures, and smoldering scorn. Our tortured contemporary education is a logical outgrowth of its origin story. The very first educational edicts for Black people in this country were the nineteenth-century anti-literacy laws that made it illegal to learn how to read and write. Even saying that now, at the end of the first quarter of the twenty-first century, fuels a smoldering rage in me that just sits there like an internal spirit reminding me not to forget—it was against the law in America for Black people to learn to read and write. From then, through the seasons of rabid de jure school segregation, and ongoing de facto segregation today in places like Atlanta, where I live, and Brooklyn, where I'm from, the assault is ever-present. In some ways, it feels definitional to American education. That is how we recognize it—a perennial struggle to exorcise the ghosts of race that are embedded in its DNA, in the nearly fifty-five million individual students, teachers, staff, and district leaders that comprise American public education, and the history that has brought them all together.

The national milestone introduces a convergence of technology and racial backlash in education that is unique to this moment in time. Yes, tech tools have always been presented as catalytic—as tools that will push education ahead. Give everyone a calculator, and learning will be catalyzed. Give everyone a tablet, and learning will be catalyzed. Give everyone a laptop, and learning will be catalyzed. By now, it is fair to say, no one is convinced that any of those allegedly catalytic technologies have materially changed the outcomes of American education. Despite the claims and clamor, well into the twenty-first century, less than 40 percent of American twelfth graders are proficient in basic literacy.[1] And the racial disparities are chronic. But now, the ferocity of regressive racial animus, coupled with the blinding enthusiasm around AI, make their confluence on the nation's big birthday special.

I have to admit: It is a fascinating time to be an American and be a part of all this, as enraging and terrifying as it is. To be here, in and of this country when it is at war with its founding principles, is an opportunity to be a part of a real battle for its future. For people of my generation, this is a first. It is an opportunity to be in the comprehensive fight to protect and defend the founding principles, which, if adhered to, can, in fact, make this nation great. The milestone is inviting us to pause and look back at our shared history and reflect on who we are, who we have allowed ourselves to become, and more importantly, determine who we will be. It is demanding of us conviction and courage, the extent of which we have not had to consider in this, the most recent fifth of the American story. What will America be in its next season? To be clear, we're experiencing a shocking return of racial animus, with leadership and language welcoming back American barbarism—one option for an answer to the question: *Who will we be?*

In this season of America, the stakes could not be higher because we are at a junction in the national road. Who we will be is surely determined by what we know of who we are. Using old terms, the idea of a first-world nation, a high-income nation, a developed nation, a G7 nation—the United States of America—banning and burning books in the twenty-first century because they are allegedly un-American is hard to come to terms with. It brings us eerily back to the German book burnings of the 1930s, when Nazis burned books they determined were un-German. And it is welcoming similar behavior. America is becoming a country where masked paramilitary of the state shoot and kill American citizens, kidnap people off the streets and from their homes and send them to clandestine torture camps. We are becoming a nation providing the weapons for and watching the wholesale annihilation and displacement of people, even after we've been taught: Never forget, never again. The attempts to gut education of so-called Diversity, Equity, and Inclusion (DEI), so-called Critical Race Theory (CRT), and so-called Woke principles are attempts to narrow the scope of what we know about who we are and, consequently, narrow the scope of who we can be. All of that is a hardly veiled attempt to smother the historic abuse of Black people and others in this country. All of that is a hardly veiled attempt to render invisible and absolve the guilty party and pretend the nation doesn't have violent and racist instincts. Education, therefore, is the lifeforce of this season of struggle. What we know is the cornerstone of who we are and will be. How we know, and the technology to facilitate knowing, are only ancillary to that basic fact.

This struggle for national definition and the battle over the content and purpose of education are colliding with the arrival of AI in education. By now, the superlatives used to describe AI are endless—both as a path to a new day, and as a path to doomsday. It is still unclear where this will all settle. The list of descriptors of AI is so long it doesn't warrant new attempts to capture its potential impact or gravity. In education, AI-enabled tools are already transforming what is taught and how. Whole schools are emerging that are AI-based,[2] and virtual AI tutors[3] are allegedly personalizing instruction and making teaching and learning more efficient and effective. Students are using AI tools to do more with mere prompts and less instruction. It can all be considered revolutionary, as the Big Tech and edtech companies that make the tools suggest it is. Perhaps. But what do these tools do to change the fundamental structure of American education? What do they do to break the perennial correlation between race and academic achievement, or race and the quality of school students attend? What does a highly efficient AI-enabled school operating in a state that has banned the teaching of Advanced Placement (AP) African American Studies offer efficiency in? AI is helping those students be efficient at learning what? How does the application of AI-enabled tools affect educational outcomes in schools that are profoundly segregated? When the purveyors of these technologies claim that they will close gaps and improve the educational experience for all students, what does that really mean?

The instinct in the tech sphere is to suggest that it is not within the purview of Big Tech or the edtech companies to influence the foundational structure of American education. Rather, their role in education is to offer tools to make it better. However, their claims of revolutionizing teaching and learning, of unlocking the potential of education, and of transforming the institution of school suggest otherwise. The cynic would say their aims are financial, and the outcomes in education are merely incidental to that end. Regardless, we cannot have it both ways—claiming to revolutionize education but doing nothing about the core injustices that warrant revolt in education. That is how the awe of AI is luring our attention away from the slow and steady assault by America's race demons. The enthusiasm, the words, and the clamor are a veil for vacuous neglect of the core justice fight at this critical milestone in American history.

More specifically, this collision of AI in education and the resurgence of race demons makes for a precarious time in science, technology, engineering, mathematics, and computing (STEM) education. For the better part of the first quarter of this century, the rise of Big Tech has made computer science (CS), and computing education more broadly, among the most popular subjects in American higher education. The rise of the FAANGs—Facebook, Apple, Amazon, Netflix, and Google—reconfigured the imagination of American young people. Now, Nvidia has taken that allure to another level. While that enthusiasm is beginning to wane, in some ways because of AI, over the first quarter of this century, Big Tech solidified the equivalence of tech skills, wealth, and power. It produced men who would start their own personal space programs. It allowed mid-level employees to become millionaires. It enabled the manifest destiny of a generation of gentrifiers. The technologies became verbs in our lives. The allure was extraordinary. It was a pathway to the new working rich, a new gilded class in America. Computer science departments at universities across the country saw record enrollments. At the nation's most elite universities, CS departments became impossibly selective. Only the nation's absolute top students were eligible for consideration, let alone admittance. In the K–12 world, STEM camps, STEM schools, STEM certifications, STEM students, and STEM curricula became the mileposts against which educational progress and value were measured. Parents of school-age children over the last several years can attest. The child studying English was met with loving lament, while the one pursuing computer science at Stanford University was the flag post of family pride. Lurking behind that social phenomenon was the quietly growing partition between those who have access to that world and those who do not—those who have access to power and those who do not.

Coincident with the rise of Big Tech and our arrival at this national milestone has been the slow and steady attack on affirmative action—the ability of higher education institutions to consider race in admissions. In this era of dismissal of justice in favor of shiny new tech things, the attack on affirmative action has found safe harbor and proved to be successful. The Supreme Court decision in the *Students for Fair Admissions v. Harvard* case ending affirmative action in 2023 stands as a symbolic marker of the ideology underpinning the efforts to make America "great" and put America "first." In this new world, attempts to acknowledge and right past wrongs are summarily dismissed. It is worth repeating that prior to 1964, it was legal in America to use the singular characteristic of being Black to deny access to all manner of American things, including higher education institutions. For perspective, members of the first free generation of Black people like me didn't get to college until the 1980s and early 1990s. The resistance was immediate, and the attack on affirmative action was already well underway.

A Black child born in 1964 didn't get to college until 1982. I started college in 1990. Considering that, another significant collision has arisen. The overturn of affirmative action, one of the most significant legislative efforts to right American wrongs in higher education, has immediately restricted access for Black students to the most elite institutions in the country. That is happening while the most advanced education and research in computing and artificial intelligence are concentrated at those very institutions. Those top institutions—the Massachusetts Institute of Technology, Carnegie Mellon University, Stanford University, University of California, Berkeley, and Georgia Institute of Technology, among others—are the places where the frontier of AI and its possibilities are unfolding. They are also the places where the door is closing on Black people. AI is among the most defining technological markers of our time, and Black people are being blocked from sitting at the table of its development. Blocked access to those institutions is equivalent to blocked access to the frontier of the tech sector because they are intimately intertwined. The people of one come from the other. The funding from one goes to the other. It is a kind of intellectual and financial slurry that bonds them together. Looking back in history, it is not a stretch to see a parallel—in order to keep Black people from accessing power and articulating freedom, the anti-literacy laws blocked us from learning to read and write. In order to keep Black people from accessing power and articulating freedom in a world driven by AI, the anti-affirmative action laws block us from the education necessary to unlock its power. Whether by design or coincidence, the outcome is the same. Black people are facing a towering struggle to gain access to the frontier of American technical power. This, at a time when the resistance to Black progress is not just the chronic low expectations of Black people born in the country's origin. The resistance is an ignited, unmasked, vengeful assault of a kind my generation has never experienced. Said another way, Black people are facing the threat of being recaptured into slavery in another form: digital slavery.

Of course, the top schools do not have a monopoly on AI education and research. Equally, the Big Tech companies and their offshoots do not have a monopoly on research and development of AI algorithmic models, products, and regulatory structures. But to think that they are not at the epicenter of the intellectual engine, wealth, and power base of this technological era is naïve. For Black people to be locked out of the generative conclave of this form of education and the inner circles of corporate research and development on artificial intelligence is to be locked out of the center of contemporary power. One American jingoistic tenet that will surely be lionized on the nation's 250th anniversary is that if you're not in power, you're subject to it. For Black people, being subject to American power is cause for nightmares.

In the lead-up to the presidential election of 2024, then-candidate Donald Trump declared that Haitians eat dogs. Haitians are Black people. More importantly, they are the only Black people to successfully revolt against European enslavers in the Western Hemisphere. In 1804, Haiti became the first free, independent Black republic in the West. Toussaint L'Ouverture defeated Napoleon Bonaparte, and Haitians have been paying for that insult ever since. They are a profoundly symbolic group of Black people, to Black people. What I heard him say from the podium in front of America is that this special group of maligned and heroic Black people eat dogs. The bile and contempt were dripping from elongated words. Subsequently, he declared that illegal immigrants are vying for "Black jobs." Those jobs in the framing of his argument are low-end, menial jobs that no one else wants to do—the kind of jobs that are beneath the dignity of those he views as legitimate white Americans. Then, on inauguration day, the captains of the tech industry were at his flank. Their presence, a tacit alignment with his thesis that Black people are bottom dwellers in American life. How could I, given the generation I am of, and the sensibilities wrought of the timing of my life, interpret it otherwise? The leaders of the AI state, the cardinals of the conclave of this new AI world, stood by and offered pardon and license for my public ridicule. The law, in overturning affirmative action, closed the door to the conclave, and the leaders of the industry locked it from the inside.

Taken together, the inability of public education to prepare Black students, coupled with the enthusiastic locking of doors to the technical frontier, is the outline of the challenge ahead. In a season of national reflection, from that point of view, digital slavery is not hysterical. History remains close. What makes this season different for my generation is the unmitigated public racial scorn coincident with the fantastic technical potential of AI. I've never seen a sign that says "For Whites Only," but I can ask a chatbot, "What about America made it okay to have such signs?" What the Big Tech titans and edtech companies are offering Black students is AI in education. That is materially different than education for AI. The former is the use of tools that are AI-enabled to help students do small things wrapped in big claims. The latter is the high-quality, rigorous education that prepares students for advanced technical fluency. It is the path to the very institutions at the frontier where the doors have been summarily closed. It is the path that includes algebra in the seventh or eighth grade, that includes Advanced Placement Calculus BC or Computer Science, that includes reading literature deeply and broadly, the path that includes rigorous, methodical, stepwise development of imagination, disciplined mental logic, and variable fluency. In the context of an America where the president says that Black people eat dogs and are destined for Black jobs, there is such a thing as Black education, and it can be enabled by AI in education. It is the kind of education that renders students prepared for a life of servitude, a life that we've known before. It is commensurate with the idea that we are a barbarous people whose ambition ought to be limited to Black jobs. It is the kind of education that, if we look back in America, we remember. That is the design of digital slavery. Go forth, you Black high school or college graduate, with fluency in tools, but not algebra. Go forth with an AI certificate that may or may not be worth anything. Follow the road to the field, not to the house. Go forth and be a rideshare driver or food app delivery person with no insurance and no benefits, a digital sharecropper of sorts, and when the algorithms from inside the conclave at the AI frontier render you obsolete, Godspeed.

This is not an argument of fatalism, rather an homage to the sanctity of memory. We've been groomed in this century to seek bulleted recommendations, the seven steps to swift transformation—of anything, from weight loss to advanced manufacturing. We've become so solution-oriented that accurately identifying the parameters of a complex problem is becoming a dying art. This challenge doesn't lend itself to swift transformation or an obvious solution. The average tenure of an urban school district superintendent in the United States is now 2.7 years; for Black women superintendents, it is 18 months.[4] That isn't long enough for a child to get from first to third grade. Transformation under what consistent leadership? The claims that AI in education and AI literacy will be transformative to American education and will elevate Black students and close gaps are absent memory. Only an explicit, committed fight for educational justice can do that. Those making the claims are not looking back, nor are those accepting them without critique. When the tech leaders professing to offer help, to offer AI in education, are the very ones standing by as we're being publicly damned, there is an inherent hypocrisy at hand. Why would I trust their claims, or their professed interest in my well-being? That hypocrisy has long American roots too. Given the 250th national milestone, mention of the Constitution is relevant. Within that very document, the contradictions concerning Black people are spelled out in the original language that originalists are so devoted to. We are the "three fifths of all other Persons," in a land where all men are created equal. In the era of George Floyd, when the Big Tech commitments to racial equity and justice were raining down like heavenly manna, in some circles, there was a sigh of relief—finally, the most powerful agents of our time were taking note and committing to racial equity and justice. One American season later, the race demons raged and all of those commitments are dropped. The language, the words, the people, the sentiment erased, the websites vanished, the funds dried, and the true spirit of self-interest revealed. The question was never: What will happen to George Floyd if I do not stop to fight on his behalf? The question all along was: What will happen to me if I don't appear as if I am? Memory is the anchor of discernment and offers clarity of sight to know which are the solid rocks upon which to stand. The glittering claims of what AI in education will do, offered by those who abandoned us in our hour of need, are not those rocks.

The Sankofa bird symbolizes the importance of remembrance. The word *sankofa* is Twi, one of the Akan languages. It means, "go back and get it." More importantly, it is tied to an Akan saying, *Sɛ wo werɛ fi na wosan kɔfa a, yɛnkyi*—it is not taboo to go back and fetch what you forgot. There is nuance there. Remembering is important, yes. But not being ashamed to go back and search for what you've forgotten is different. It is a more powerful offering. It is an offering lined with grace. It says that it is okay to not know once you're committed to knowing. The effects of American education on Black people are such that there is so much we don't know. In the long arc of the American story, we know the nation's lessons in the scars and stripes on our skin. But the structure of American education has rendered it such that Americans, all of us, don't know the full lessons of the American story. The current efforts to strip education of those lessons are an explicit attempt to keep it that way. While the AI effort in education is here to stay, our ability to look back and fetch lessons from the past is important for the potential of AI in education to be justly realized. That ability helps us see patterns and connect historic plots in the dark. It helps us position this new technology for what it is and not allow it to become a substitute for what it is not. Our ability to look back is necessary for us to calibrate the claims surrounding AI in education against the reality of what we know and what we've forgotten but must remember. Not only is memory a lens for the clarity of sight; it is an anchor for hope. As the Sankofa bird looks back from its perch at this national milestone, it is looking back from a perch while free for the first time. That is evidence we have succeeded. We're unwilling to be erased. Though my father was born a subject of the Crown in form, he was never subjugated in fact. While the risks of digital slavery lie ahead, the ashes of actual slavery lie behind. It didn't win. Hence, remembrance is not just *of* resistance; it *is* resistance.

The Bullwhip's Algorithm: Artificial Intelligence and America's Racial Caste System is about the convergence of memory and the promises of AI in education. It is an attempt to pause and examine the actual foundation on top of which the grand claims of AI's potential in education are being built. It is an attempt to look without blinking or misspeaking at the truth of the American hour. The race demons are back, daring us to call them by name and threatening us to forget the terror of their trespasses. All the while, the captains of AI are positioning a new narrative. They suggest that AI will fundamentally transform the experience of education. But they are doing it at the same time that they have turned their back on the very people for whom education has ceased to be transformative.

The first section, "Artificial Education, Real Intelligence," addresses the challenge of gutting American education of the American story. The creation of artificial education made more efficient by AI can create a national hallucination of who we are. The parallel ongoing emphasis on STEM education can be a distraction to that phenomenon and create a set of false expectations about how equitable outcomes are achieved. The second section, "Inverse Correlation: Power and Courage," as the title implies, considers the relationship between the most powerful actors on the technical stage at this historic national milestone and their courage, or lack thereof, to defend what they presumably know is just and right. The world has seen silent accomplices to grave injustice before, using exceptional positive highlights as an excuse to turn away from ghastly patterns. As such, the importance of having second sight. The third section, "21st Century, Second Quarter," identifies the difference between the philosophy of technological supremacy and the foundational philosophy of justice. Sadly, the two are at odds with each other. The conflict between them determines whether, as a nation, we seek highly skilled but uneducated workers or a population of educated citizens who can also work. Educated citizens are the defenders of democracy. Mere workers, both skilled and unskilled, are just a workforce, not a nation of citizens. Lastly, "Must Be Freedom" is a nod to justice. In the midst of all the noise and clamoring for shiny things to do a lot and mean even more, there is justice. It stands above the noise and the masquerade. It only bows to freedom. That is its only master. In a time of remembrance for America, it is freedom that sits behind the altar, and justice is communion.

From my memory to yours, with love.

PART I

ARTIFICIAL EDUCATION, REAL INTELLIGENCE

CHAPTER 1

HALLUCINATION AND SEGREGATION

In 1963, as the Civil Rights Movement was reaching its climax, James Baldwin addressed an audience of schoolteachers in Philadelphia. He said we "must understand that in an attempt to correct so many generations of bad faith and cruelty, when it is operating not only in the classroom but in society, you will meet the most fantastic, the most brutal, and the most determined resistance. There is no point pretending that this won't happen."[1] In 2024, more than 60 years later, the State Superintendent of Schools in Georgia banned the teaching of Advanced Placement African American Studies in public high schools across the state.[2] The sentence itself is hard to contemplate. The United States is 250 years old in 2026, and states are banning the ability to look back—a critical feature of a self-aware and well-informed modern democratic state. Baldwin's resistance is real.

While Georgia and several other states are banning the teaching of history, the future is unfolding at warp speed. In this same season of criminalizing historical knowledge, AI has taken the imagination of the world by storm. From AI-generated music raising questions about who the actual musician is, to the Screen Actors Guild strike raising questions about whether we need human actors for movies anymore, it has upended much of our confidence in what we know as a society. It has introduced wonderings, the questions for which we haven't yet been able to formulate. Its implications are so vast and uncertain the Biden administration in 2023 issued an executive order on the use of artificial intelligence. It says in part, "In the end, AI reflects the principles of the people who build it, the people who use it, and the data upon which it is built. I firmly believe that the power of our ideals; the foundations of our society; and the creativity, diversity, and decency of our people are the reasons that America thrived in past eras of rapid change."[3] These are the kinds of lofty ideals that are the bedrock of the American story. They rest on confidence in an idealized and ambitious version of the American people. The coincidence of these phenomena—the rise of AI, the attempted erasure of history, and a utopic belief in the nobility of the American people—has pushed the country to a dangerous precipice unique to our time.

Building a digital future uninformed by actual history is a dangerous thing to do. A digital future is one where the intersection of human agency and technology is blurred. It is the world we're moving into—one where technology's penetration lies not just in how we use it but in how it learns from us to generate what we think we know. It can create a kind of hallucinogenic utopia that, in America, hides its racial demons. The foundations of our society—and the diversity and decency of our people that President Biden called on—have the potential to disappoint. The very history being banned articulates those disappointments and helps arm the society to prevent them from happening again. Accurate history is a necessary input to the construction of a digital future that will surely be heavily influenced by AI. Having such strong belief in the virtues of the American people, especially in a time of such national flux, is naïve. Black people, the subjects of the banned history, know better. It is the same American people whose decency and sense of national pride are fueling efforts to ban books and have teaching Black history made illegal. And they are doing so more than 60 years after the passage of the Civil Rights Act and more than 160 years after the signing of the Emancipation Proclamation.

The decision to ban African American history in Georgia is not the whimsical twitch of an errant racist deferring to an archaic antebellum American law. It is the decision of a sitting state superintendent of schools backed by a 2022 Georgia state law, HB 1084—the Protect Students First Act, commonly known as the Divisive Concepts Law. It is evidence of the resistance James Baldwin warned of. There is no point pretending that it is not. We have seen this before.

To be an American and to be Black has always been a fraught proposition. More than a hundred years ago, W. E. B. Du Bois shared in his essay

Strivings of the Negro People, "One ever feels his two-ness,—an American, a Negro; two souls, two thoughts, two unreconciled strivings; two warring ideals in one dark body."[4] In the American slave days, white people banned Black people from learning to read with anti-literacy laws. It was an offense in those times punishable by fines to white people caught teaching Black people to read, and by whipping or death for Black people trying to learn. The universal understanding then, as now, is that education is an exalted power. Unlocking the keys to knowledge through literacy was a direct path to resistance, to the calling out of hypocrisy, to a more articulated call for justice, to a more robust demand for citizenship and human dignity. Banning the teaching of history today is analogous to the anti-literacy laws then. It keeps people unaware and unable to correct past wrongs. It blinds us to who we are. Even knowing that there was a time in America when white American citizens passed laws to prevent the Black people they owned from learning to read is part of the historical record of the United States.

On one hand, there is the twoness of being Black. The challenge of having to reconcile the idea of being an American with the reality of being Black in America. Those ideals in the American story were born at odds with each other and remain in conflict. On the other, there is the twoness of being white: facing the challenge of being wrapped in an American flag tailor-made to fit a white body, while an ancestral record of unspeakable barbarism and inhumanity looms in the background. For Black people, it is a conflict with the country. For white people, it is a conflict with themselves. The relationship between these two distinctly American conflicts—the struggle to triumph and forgive and the struggle to repent and reconcile—is in many ways the heart of the American story. It is an intimate, painful, incremental, and inescapable relationship. It is the relationship that lets us know that African American history is American history. This is the relationship where, in fact, there is no daylight between the two bodies of history, the two strivings to understand, and the shared interests in a common outcome. Neither of us can understand or appreciate the American story without fully understanding ourselves—who we are and who we have been relative to each other.

Banning African American history, therefore, is banning American history. It is banning the teaching, learning, and understanding of this complicated relationship born of America's Original Sin. Banning this knowledge of American history will fail to arm generations of students with the contextual understanding they will need to withstand cynicism and keep fighting toward a more perfect union. Contrary to the name of the law, it will leave students unprotected from the nation's racial demons. They will not be able to recognize racist movements cloaked in nationalist populism or right-wing ultra-conservatism. They mightn't see that Christian faith has been used both to justify slavery and as a freedom light to keep on keeping on. They won't know how effective social protest movements helped push the nation forward, or how resistant America was to having Black people be recognized as people, let alone acknowledging African American history as American history. They won't have a sense of the scope and viciousness of white American racism or the enormity of the challenge of exorcising it from the nation's DNA. They won't appreciate how willing Southern states were to destroy the entire country rather than concede the human and moral abomination of American slavery. And they won't understand the significance of this time of pendular reversal, where generational progress is being upended at the constitutional level.

Ultimately, banning the teaching of history undermines the strength of democracy. The whole point of education in a democratic society is to build an informed citizenry that is critical and courageous and is its own check and balance against tyrannical and racist demagogues. Demagoguery requires a certain mix of fear, awe, and ignorance that well-educated citizens don't yield to. All of these august claims are true. They are the substance of grand rhetoric because they are designed to inspire resistance to oppression in societies. But they are also serious lessons learned from history. Autocratic national populists and murderous, racist colonial regimes of the nineteenth and twentieth centuries all sought to suppress education and artificially shape the contours of history. They did so to support the ideology that fueled their particular brand of oppression.

Controlling what people know—and don't know—is an effective means of controlling people themselves. Violence is often a complement to selective education. The two serve to reinforce each other and breed both ignorance and fear—the substrate of oppression. In the American case, both are now present. The effort to ban Black history is here. The Capitol of the United States, the temple of our democratic faith, was stormed by white nationalists and populists who killed and injured scores of police officers trying to defend it. They set up hanging gallows and threatened to lynch the vice president of the United States and kill the Speaker of the House of Representatives. The society's demons are clearly restless. The reshaping of history and the insertion of violence are here. The ingredients are present, and the restive racial demons are on the hunt for new ways to oppress.

We are in a time when state officials feel free to openly denigrate and scorn Black people. In the tradition of Governor Wallace of Alabama and Governors Talmadge and Maddox in Georgia, they feel emboldened to say in public that Black people's history—and by extension, Black people—are irrelevant. Their position is that not only is it not worth teaching, but that it is dangerous and therefore ought to be banned in public education. In 2023, Governor DeSantis in Florida approved into law a state educational standard impacting millions of public school students that asserts, "[S]laves developed skills, which in some instances, could be applied for their personal benefit."[5] Some of the comparative what-ifs are inconceivable. What if the Florida governor here, deep into the twenty-first century, were to authorize a state education standard that says, "White Americans who gathered their families and children with merriment to watch and celebrate the burning and dismemberment of Black human beings suffered from a psychosis that needs to be exorcised for the benefit of the country"? The mere thought of that is incomprehensible in modern America. Yet the state superintendent of schools has banned the teaching of African American history in Georgia, claiming it is divisive.

There are, today, formal state-sanctioned educational standards in the United States that teach children that slavery was beneficial to Black people. Banning the actual history and constructing shockingly offensive and nonsensical education standards is a process of rewriting history, of controlling what people know and don't know. That is the process that the ban is an indicator of. The moral attack and injury are evidence of the resistance that James Baldwin foresaw. In this case, it is an attempt to erase American historical data. It is the manufacture of artificial education to undermine real intelligence. It is likely designed to absolve the white American psyche of having to confront its ghosts and repent, but more insidiously to signal to the nationalist soldiers that the resistance is on and the pendular swing toward a more pluralistic nation and more perfect union is under grave threat.

The surge of these racial demons and the risks they present are happening in a season of fantastic technological progress. We are in an AI season, a season of rapid change. Advanced computational algorithms are combing through data on the internet to discern what they determine to be the truth. Apart from the technical details of what is happening, the relationship between the data on the internet in the form of large language models and educational standards is direct. If a student prompts one of the new AI-enabled tools: "Was American slavery good or bad?" It may return: "It depends; sometimes the overproduction of cotton by slaves lowered the price, but the slaves themselves acquired skills that were personally beneficial." Banning accurate history and twisting the facts has a direct effect on the output of artificially intelligent tools. Their so-called intelligence is derived from data on the internet, and that data is born of the information people put on it. That includes legally sanctioned state standards of education.

That is why the coincidence of these two phenomena is so dangerous. Skewing the data, erasing and twisting history just as the engines of AI are revving up, increases the risk of fossilizing false history. And the AI race is certainly on. The engines are revving. The enhancement of products and the creation of new ones powered by AI capabilities is the foundation of current Big Tech competition. It is paradigm-shifting in the tech space. It is so transformational that it will inevitably lead to the rise of a new generation of Big Tech and the extinction or transformation of some of the current giants. The question—Who will lead the AI race?—is being asked of venture capitalists, futurists, journalists, the market, the government, and captains of the industry itself. The heads of Google, Anthropic, Microsoft, and OpenAI, for example, have emerged as modern oracles of our technical future. Despite the growing tech backlash, we, the general public, are looking to them with trepidation and awe, wondering how AI will transform our lives. How will it transform the way we work, the nature of our health care, the way adults and children learn, whether we will have to drive cars anymore, how we will tell the difference between augmented reality and real reality, how wars will be executed, and who or what answers the phone when we call for help? Ultimately, the question we're asking is, how much of our human agency ought we surrender to the technologies these companies build? Governments are asking the same questions. Because the technologies are so advanced, governments are forced to turn to the technology companies to ask them how to regulate the technologies that they themselves build.

AI has taken on such a towering position in our modern imagination that we have given it human traits. What will AI do? What will AI say? Can AI make it better? Or will AI make it worse? Will AI replace you? And existentially, what does AI think? Notwithstanding the formal and formidable body of technical knowledge and computer science AI and machine learning rest on, they are now in the public sphere with a cultural persona. Far from the academic laboratories at the nation's most elite institutions of computing—UC Berkeley, MIT, Stanford, Carnegie Mellon, Georgia Tech, and others—AI has become a social creature. In some ways, we have bestowed agency upon it. We've even given it a fancy title: Agentic AI. In a very reductionist formulation of the challenge, we are grappling with our own agency. How much do we surrender to AI? We are navigating our relationship with AI, as a proxy for technology more broadly, and trying to find where to plant the flag of human agency. Because we are so unsure of where to plant our flag in such a rapidly developing technical environment, global leaders like President Biden assert that wherever we plant the flag, it will be informed by the tradition of American decency—for Black people, a dubious concept at best.

Generative AI is the section of this new technological frontier that is so amazing at the moment. It is the apparent generation of knowledge that we are so in awe of. It is one of the first massively accessible computational tools that has lured us into thinking that machines are, in fact, intelligent. The history of AI and machine learning is long, and it is its own body of knowledge that, for years, was largely the purview of the advanced computing research community. When IBM's Deep Blue supercomputer beat a world chess grandmaster in 1997, the technical accomplishment was extraordinary.[6] The significance of that milestone, however, was largely a passing marvel to the public on the news. It was inaccessible to the public and wasn't something that affected our daily lives in any obvious way. Up to that point, and in the years since, the progress of technology and computational ability has continued at an increasing rate but hasn't dramatically affected the public's view about the purported intelligence of machines. Even the near ubiquitous presence of technology in our lives through social media, digital commerce, and smart tools like Siri and Alexa has likely had only incremental effects on our collective sense of artificial intelligence. That, of course, is an empirical question for the techno-sociologists and anthropologists. What do we mean by intelligence and human agency? Those are the existential questions that will swirl in the academy for years. In some ways, AI has been a boon for academic philosophers and technology futurists.

There is no question, however, that OpenAI's release of ChatGPT in November 2022 was pivotal in people's relationship with technology. It was a shock. Within about sixty days of its release, more than 100 million people had downloaded the application.[7] By now, it receives multiple billions of visits per month, but it is the generative component that has left us in awe. Here, too, there is a massive body of computational accomplishments behind the technology that require delving into, but on the surface, the tool has been miraculously transformative. You can ask it in the prompt, as I just did, to "write a three-paragraph essay about the significance of the Haitian Revolution." Following the flashing dots, which we now know are designed to bait our breath, it begins:

> The Haitian Revolution (1791–1804) was a watershed moment in world history, profoundly affecting the course of human rights, colonialism, and slavery. It began as a slave revolt in the French colony of Saint-Domingue and escalated into a full-fledged revolution that resulted in the establishment of Haiti as the first independent Black republic. The revolution was the first successful slave uprising, demonstrating the possibility of enslaved people overthrowing a colonial power. The triumph of the Haitian Revolution became a symbol of resilience and hope for enslaved people across the Americas, profoundly impacting abolitionist movements and spreading fear among colonial powers.

Even as I write this now, having studied mechanical engineering at UC Berkeley and gotten a PhD from Georgia Tech, I realize that I went through elementary and high school before computers were really a thing. I am one of the last generations that had to use the library in its old form. I had a class at the Brooklyn Public Library to learn the Dewey Decimal System so I could find individual books in that cavernous building. I then had to use my library card to check them out, carry them home, read them, take notes, and then synthesize the information to determine the significance of the Haitian Revolution. I had to write essays in long form before we got a typewriter and ultimately a word processor. Sometimes when I pause for a moment in front of ChatGPT today, I find myself awestruck. I feel like people must have felt after having installed the first light bulb in their home. Flip a switch on the wall, and there is light. Voilà! No need for fire. It is a technological marvel of the new world. Now, on my phone, I can give AI a simple prompt, and it gives me back a fully formed, coherent, structured essay in precisely the manner I asked, in seconds. Of course, the applications in generative imagery, the generation of voices and music, and the ability to learn are all equally mind-boggling. Professors and teachers have been quibbling about their ability to sniff out whether written works are those of their students or not. It is only a matter of time, as the technology gets better and develops greater acuity, before that debate will be rendered moot. Deepfakes are already extremely difficult to sniff out. It is a matter of debate for sure, but the ability of a technical tool to gather information and generate coherent syntheses with a particular point of view is a cornerstone of what many in the general public consider intelligence.

In many ways, the generative characteristics of AI have brought a beautiful and sweeping sense of wonder to our time. It has given us something to behold. Undoubtedly, think tanks and research centers are in overdrive assessing and predicting the impacts on education, the workforce, warfare, and all facets of human engagement. There has always been a parallel study of the impacts of technology on society and the nature of the society that produces the technology. At many universities, there are schools of "Science, Technology, and Society" dedicated to the interplay between our technological products and their impacts on the societies in which we live. There is nothing new about that interaction. The difference that makes the current AI moment special is its ubiquity. Hundreds of millions of people are experiencing it. We don't need academic studies to appreciate the potential impact. This is citizen science on a global scale. Huge populations of people are experiencing for themselves its sheer power and transformative potential.

School systems, for example, are grappling with how to build policies around effective pedagogical use of AI-enhanced tech tools. They are considering what to teach and how, but also how to distinguish what students really know from what AI has generated for them. At the same time, students are experimenting, and there are millions and millions of them. We are all experimenting. The use of the tools themselves is becoming its own literacy. School systems and businesses are trying to determine what AI literacy means and, hence, what AI fluency is. What is it that people need to know how to do to be effective in their human endeavors? Because of that direct individual experience, the sense of wonder, awe, and appreciation of the transformative potential in every sector is more real. Can my job be replaced? Will my profession be rendered irrelevant? How can my imagination be enhanced? What is the future of school? These are real questions on the minds of hundreds of millions of people. This level of technical penetration is arguably more pervasive than previous ones and is giving us more pause. Are these machines intelligent? Is our intelligence itself on the precipice of obsoletion? It is not a passing marvel that we are being told about. It is a technological leap forward that we are experiencing.

That widespread sense of wonder, or at least curiosity, leads to the obvious questions: How does AI know? And how does it synthesize? These kinds of questions have always swirled about computers, and in some ways, we have become comfortable not knowing exactly how all these technologies really work. It is more important to the users *that* they work than *how* they work. The generative ability of these new tools, however, has reintroduced the questions. How does ChatGPT know what I mean by the *significance* of the Haitian Revolution? Significance to whom? Or to what body of thinking? And how does it so quickly not only gather the relevant facts but also position them in a historical context that gives them contemporary applicability and predictive relevance for the future? And why does it not conclude, to the contrary, that the significance of the Haitian Revolution was to catalyze colonial regimes and their descendants to build systems to forever punish and prevent Black republics from thriving? It might not have an opinion, but it is certainly presenting a particular point of view. While the merit of the question is relevant to the current time we're living in, how ChatGPT and Google's Gemini, for example, gather information and synthesize it is at the heart of the question: *How does it know*?

The algorithmic infrastructure behind generative AI is the purview of technical experts. There are tomes of literature and research dedicated to the computational and technical underpinnings of artificial intelligence.

The layman's understanding is that AI-enabled tools are churning through massive amounts of data on the internet. Large language models (LLMs) is the term for the tools that combine sets of data. The challenge with this process is that the data from which tools like ChatGPT, Claude, and Gemini are learning are merely data; they are not gospel.

The syntheses that these tools generate are only as good as the data upon which they are based. LLMs make use of the vast bodies of literature, research papers, newspapers, text, and images of all sorts that exist on the internet. Hence the problem. There are no rules that guide the veracity of information on the internet. All manner of claims of truth exist, as we are fully aware in this era of fake news and alternative facts. Because something is a search result on the internet or exists in the digital space does not make it true. So, when ChatGPT and other tools like it are combing through data to arrive at the kinds of answers to questions—like, "What is the significance of the Haitian Revolution?"—they get it wrong from time to time. It is called AI hallucination.

Hallucinations, as the name implies, are instances when generative AI results are incorrect representations of the truth. That can be in a host of forms, the most obvious being images and text. There are many other kinds of mistakes generative AI can make that are far more dangerous. This is particularly true when AI tools are used to build correlations between people's images and potential criminality. The mental relationship between images of Black people and crime is woven deeply into the American psyche, for example. That relationship is reflected in the data that AI tools use, which can lead them to generate and reinforce that relationship. Safiya Noble's *Algorithms of Oppression* and Joy Buolamwini's *Unmasking AI* capture the complexity and danger of the data problem.[8] The tools hallucinate. They present incorrect results either because they misinterpret the data used to generate the results, or the data itself is inaccurate or just missing.

The hallucinations of generative AI tools are caused by bad, biased, or incomplete data. There is an analogy to historical data in schools. Having bad history leads to a hallucination of what America is. Enter the state superintendent of Georgia banning Black history in schools and the governor of Florida authorizing an educational standard teaching that slavery was beneficial to Black people. There is a hallucinogenic parallel that ties the technological leap forward we're experiencing to the pendular reversal happening at the same time. Banning and contorting history is manipulating the data. It is building an inaccurate dataset for learning that will yield young people who generate a false understanding and synthesis of history. The danger here is that the technological leap we're experiencing is so awesome and existential that it has the potential to lure our attention away from the growing resistance of American demons, from the reversal of the pendulum. The erasure of Black history and the sanitizing of white history make for bad, biased, and incomplete American history. Embedding inaccurate history into educational standards leads, for example, to several million American whites clamoring for a return to an era when America was "great." As Langston Hughes suggested, "I, too, sing America. I am the darker brother."[9] From my point of view, I am of the first generation of Black people born fully free in American history. What era are they referring to when America was great? Or perhaps it was, in fact, great, but they just don't mean for me.

Our collective grappling with AI is happening at the same time the nation is celebrating its 250th birthday. This all-consuming era of AI also overlaps with the seventy-second anniversary of the Supreme Court's *Brown v. Board of Education* decision establishing that "schools separated by race are inherently unequal." Both are historic milestones for the country. They are forcing us to reflect on who we are as a nation, who we have been, and ultimately, what kind of nation we want to be. The significance of accurate history during these milestones cannot be overstated. It is the bedrock of our reflective ability. It is the dataset we'll use to generate the synthesis of the American story as we contemplate our future together. On one hand, we are grappling with the future of human agency in a world of increasingly intelligent machines. On the other, we are grappling with our earthly history together and whether and how to confront America's ghosts.

The 250th anniversary sets the stage, and the *Brown* decision is a useful point of entry. It helps us appreciate the distance between the reality of who we are and the hallucination of who we claim to be. Because AI has taken our imagination by storm, the question of how to codify and deploy appropriate AI-related skills and AI-based tools is dominating the educational landscape. It is the next frontier of STEM—science, technology, engineering, and mathematics—education. Equitable STEM education has been a goal in education quarters for decades. Now, as we contemplate artificial intelligence in the STEM education sphere, it is forcing us to revisit reality and deconstruct some of our collective hallucinations about America.

The 1954 *Brown* decision is iconic in the civil rights portfolio. In many circles, it is considered the crown jewel of the civil rights era. It is also the subject of tomes of research on its efficacy, what it signaled for American public education, what it did to Black teachers and Black schools, and what it symbolized for the progress of the nation. By now, it is widely understood that there was massive resistance by white communities all over the country to having their children educated alongside Black children. White people did not want racial integration. Many of the pioneering Black people, who as children and young people had to brave the viciousness of white people as they tried to break the educational color barriers, are still alive. We have firsthand information and have not forgotten. Ruby Bridges, the unsuspecting little girl turned American hero, integrated the William Franz Elementary school in New Orleans, Louisiana, when she was six years old. Ironically, she was born in 1954. My parents are both older than she is. In 1960, she was the first Black student to integrate a public elementary school in the American South. She is the subject of Norman Rockwell's famous painting *The Problem We All Live With.* It depicts the child, Ruby, being escorted by federal marshals to school with a tomato splattered against a wall with the word "nigger" scrawled on it. These scenes are repeated throughout the American educational landscape during that era, from Little Rock to Boston. Mobs of white people, many of them parents of children themselves, spitting and spewing venom at Black children trying to attend school. It is well documented that many white communities opted to riot and close schools altogether rather than have Black students attend.[10] The resistance to integrated schools is reinforced with blood and acrimony and the galling hatred expressed by white Americans.

It is relevant at this moment when states are attempting to ban African American history in public schools that once sought to ban Black children themselves. Whose history is this? Is this Black history, white history, or American history? There is a classic photograph of fifteen-year-old Elizabeth Eckford, one of the Little Rock Nine, taken as she walked to enter school on her first day. The Little Rock Nine were the students who integrated Little Rock Central High School in 1957. The photo captures Elizabeth walking in the middle of a mob of white girls with federal soldiers in the background. Hazel Massery is one of the white girls screaming and overflowing with racist rage. The hatred and the venom of the white girls around her, even looking at it today, is hair-raising.

I learned about this photograph and the Little Rock Nine in middle school—Mark Twain Junior High School for the Gifted and Talented—from my larger-than-life Black eighth-grade social studies teacher, Ms. Celestine DeSaussure (Ms. D.). I was one of few Black students in the school. There were a lot of European immigrants—Italians, Russians, Greeks, and a lot of Chinese students. Ms. D. led us into that lesson through the eyes of Elizabeth. We spoke about the stunning amount of courage and composure she must have had to be able to endure that kind of hatred and physical and emotional assault. She forced us to consider how terrified Elizabeth must have been and how alone she must have felt. We were fourteen-year-olds. That was only twenty-eight years after the Little Rock Nine ordeal. Ms. D. asked us how we thought we would have handled it. She made us write essays about it. We had to read the words of the Little Rock Nine themselves. We had to watch PBS documentaries and listen to them in their own voice describe their experiences. She asked us whether we thought Elizabeth's parents were reckless or heroic for subjecting a fifteen-year-old girl to such obvious and intense emotional and physical danger.

Some of my Black classmates' families were from the South. Their grandparents and parents had to make precisely these decisions. Elizabeth's experience was inside our households and families. How would her parents have felt if she were shot, or worse, lynched? We learned about martyrdom for a just cause. Her parents surely knew that for Black people to depend on the government to protect our children was the safety of fools. Would we have done it? Ms. D. forced us to discuss forgiveness. Could we forgive Hazel specifically, or what Hazel represented generally? We also spoke about Elizabeth's mental toughness. How did she learn basic algebra, for example, under those circumstances? And what were the American history lessons they offered in a school that, until she arrived, banned Black students from attending? What did they teach white students when Black students weren't around? Ms. D forced us to consider the historical data and tangle with it.

Ms. D.'s entering the photograph through Elizabeth did two things. One is that it humanized Elizabeth. She forced us to consider the awful and complex historical context within which Elizabeth existed. And she pushed us to see ourselves as Elizabeth. Humanizing the Black American experience in formal public education was its own revolutionary victory at that time. It was part of a huge leap forward, given that in 1985 it had only been twenty-one years since the signing of the Civil Rights Act, acknowledging that Black people had to be treated equally. The other, upon reflection, is that it highlighted that we did not enter through Hazel. Elizabeth, and the Black people she represents in the photograph, had been diminished and ignored for so long in the larger American story that what Ms. D. did was Herculean. She was lifting up the Black narrative for us, eighth graders of many races and ethnicities, and placing it firmly into our formative conception of America. She was enriching our data set, making it more complete. In doing so, the other thing Ms. D. accomplished was forcing us to recognize that grappling with the humanity of Hazel is the core of the American challenge. Hazel is the problem. In some ways, the current movement to ban Advanced Placement African American Studies and the legislation against so-called divisive concepts is not to ban Elizabeth; it is to ban Hazel. That is how the unexamined and innocent Hazel becomes the hallucination. The standard treatment of segregation is passive. There was segregation, then there was Martin Luther King Jr., then it went away. *Segregation* is a noun, but *segregate* is a verb. Someone had to do it, and it is still being done. American history in the passive voice removes the burden of responsibility from Hazel. There is a guilty party.

When you enter the photograph from Elizabeth's point of view, triumphant, terrorized, courageous, resistant, poised, and crusading are the adjectives of entry. Ms. D. forced us to unpack them, grapple with them, contextualize them, put ourselves, to the extent we could push our imagination, into her shoes. Ms. D.'s point of entry through Elizabeth's eyes is part of the crown jewel narrative of *Brown*. It is the more righteous lens. It is the story that allows for the amplification of American virtue. A nation that saw it was allowing racism to tear it apart, and through the highest court in the land, and on the backs of courageous young Black people like the Little Rock Nine, sought a more righteous path to a more perfect union. It is this interpretation that leads scholars like Nikole Hannah-Jones in her essay in *The 1619 Project* to say that America owes much of what virtue it has to Black people.[11] It is the violence of the American story as seen through the eyes of the victims of its violence, not the perpetrators of its violence. That is the story of redemptive suffering.

Entering the photograph through Hazel screaming at Elizabeth offers an entirely different point of view. It is the part of history that we generally don't have access to—interrogation of the source and experience of white people frothing with racism and fury. We don't see textbooks or standards or exams that contemplate the racist rage within a young teenage white girl. I never went through an exercise, like the one Ms. D. led in our class, to consider Hazel's point of view. What was it about Elizabeth's being Black that ignited so much fury in Hazel? What strand of American values was adopted in her family's household and in her community that produced her? More broadly, what was it about white Americans that led them to be so barbarous and almost medieval with their tribalism and violence—even in response to Black children just going to school? We don't see parallel educational standards in Florida, for example, which offer that "attempts at integration, in some instances, triggered in white communities a racist rage that has revealed itself to be a permanent detriment to equitable public education and the nation's well-being." We don't interrogate the contrary view that the photo, rather than symbolizing a triumphant step toward desegregated schools, symbolized a new phase of white resistance to integration. A more complete and accurate historical record is the foundation of a more complete data set. Without it, we hallucinate.

That obfuscation of historical data allows us to accept conditions that we would not publicly declare. It is a kind of permanent ethical transgression hiding behind historical omission. In our case, in the American case, if and when we give it thought, we hang on the conclusion that the 1896 Supreme Court ruling in *Plessy v. Ferguson*, which allowed for nearly seven decades of legal apartheid in the United States, was a bad decision. The Jim Crow era that it wrought is publicly acknowledged now as a stain on the American cloth. But Hazel was a defender of it, as were thousands and thousands of other white Americans. Notwithstanding Hazel Massery's personal change of heart in her later years, we don't interrogate her in formal educational standards and settings.[12] And given the permanence of racial segregation in schools, what position do the communities and attitudes Hazel represents currently hold?

It is easy to simply reduce white communities' resistance to racism. Hazel was forced to accept Elizabeth in her school sixty-one years after *Plessy v. Ferguson* and only three years after *Brown*. They were born in the early 1940s. It is possible that their grandparents were born around the turn of the twentieth century when the *Plessy* decision was handed down, and their parents likely grew up in the 1920s and '30s. Those were the years when the American Ku Klux Klan was at its height, with millions of members across the country. Killing Black people was an American blood sport in those days. As is the case in so many dimensions of contemporary American life, because we don't aggressively address Hazel's point of view as the root problem to be deconstructed, we continue to deal with its effects. The current segregation of American public schools is such an example. In modern attempts to desegregate large urban school districts with big populations of white and Black students, the resistance of white communities has surged again—New York, Boston, San Francisco.[13] But by and large, segregated schools are the norm in modern America, and despite our unwillingness to declare that we want separate schools, we willingly accept the fact that we have separate schools and that they are unequal.

My daughter graduated from Maynard Holbrook Jackson High School in Atlanta Public Schools (APS) in 2022. In the APS system there are ten traditional public high schools. In 2024, seventy years after the *Brown* decision, seven of those schools did not have a single white student enrolled. Not by percentage; by individual students. All of the white students in the APS system attend only three of the high schools. What that means is that there are essentially seven public high schools for Black students and three for white students. Our language has evolved since the early twentieth century. Black people are no longer referred to as *colored people*; rather, we are *people of color*. But white is still white, and we still have schools for coloreds and schools for whites. Even in the schools where white students attend, they are segregated within those schools. They are concentrated in the marquee programs of International Baccalaureate (IB), Advanced Placement (AP), honors, and gifted courses. In the International Baccalaureate Diploma Programme at my daughter's school in the class of 2026, 70 percent of the students are white, and 19 percent are Black, despite the population of the school being 23 percent white and 65 percent Black.

The whole premise of the *Brown* decision was that schools intentionally separated by race are inherently unequal. That was true then, and it remains true today. While the dissolution of the Black educator class was a blow to the excellence of some Black schools in the American segregation days, in the main, the isolation, criminal underfunding, and disdain of Black schools within public education systems rendered them unequal to their white counterparts. They were separate and unequal then, coming out of the Jim Crow era leading up to the *Brown* decision, and they remain separate and unequal now. There are a host of educational indicators that education policy reformers and advocates use to determine the performance of schools that confirm this reality.

In this modern era ushered in by President Bush's No Child Left Behind Act, the American public education system cannot escape standardized testing. We use them for accountability and efficacy. In the case of schools where students are clearly coming from challenged communities, making direct comparison with those students whose lives are comfortable, enriched, and privileged doesn't make sense, so we measure comparative rates of growth—typically referred to as "gains." After all the policy contortions, complicated use of measurement and data, and grand claims of equitable paths to meritocratic outcomes, in the end, students will have to achieve mastery in foundational subjects to be competitive in certain fields. Gains are important, but they can hide the absence of mastery in segregated schools. This is where the reality of our racially segregated system confronts the platitude-laden language we use to get around the fact that we accept separate and unequal schools. Hopeful language cannot escape the simple truth. In America, with its torturous racial foundation, schools separated by race are inherently unequal.

Bob Moses was one of the heroes of the American Civil Rights Movement. He moved from Harlem to the Mississippi Delta to engage in the movement for Black voting rights. Because his origins were as a schoolteacher, he was always clear that while voting was foundational to American democracy, education was foundational for full participation. In his book *Radical Equations*, he makes the case that algebra is one of the foundational skills that Black students must have in order to participate in the tech-based economy that was gaining steam at the end of the twentieth century.[14] If we just hold his axiom for a moment, without the painful circular debate about whether algebra is as important as many people in tech fields think it is, the racial disparities in algebra performance are shocking.

Returning to my daughter's school. In 2023, 10 percent of the Black students were proficient in ninth-grade Algebra I, while 60 percent of the white students were.[15] The white students, of course, are concentrated in the more advanced classes. At the district level in Atlanta Public Schools in 2023, 12 percent of Black students and 76 percent of white students were proficient in Algebra I. This is what the unequal outcomes of racial segregation look like. Despite popular school district mantras and slogans—to empower tomorrow's leaders, to make all students college and career ready, to ensure that all students reach their highest potential, to equip all students with the skills to be a positive force for change, to prepare all students for social and economic success—in Atlanta Public Schools, in the seven high schools that are nearly 100 percent Black students, the average Algebra I proficiency rate is 4 percent. That means that four out of every hundred Black students are proficient in basic algebra. And *proficient* itself is ambiguous. While Atlanta is the example here, this pattern is national—Chicago; Los Angeles; Washington, DC; New Orleans; Oakland; Boston; Ferguson. It is all varying degrees of the same—awful. At this stage of the story, we don't need revelatory evidence to further illustrate something we already know.

This dynamic, the seemingly intractable problem of racial segregation in American public schools, is a cornerstone of American history. It is foundational to African American history and the racial hierarchy in American life. If the Little Rock Nine saga is considered African American history and not American history, this is what the Georgia State Superintendent is banning. It will ultimately rob us of understanding the depth of white resistance to racial integration in American public schools. Whether Black people wanted integrated schools is its own question that will be left unexplored without African American history courses. It is likely that what Black people wanted were good schools, not integrated schools. If the tenets of *Plessy v. Ferguson* were actually upheld and separate schools were made to be equal in fact, it may have been an easier road. It may have been an easier proposition to build good schools that serve Black students rather than subjecting generations of students like Elizabeth to the moral injury, social isolation, and constant harassment of resistance to Black presence in white schools. It may have been an easier proposition than to pretend that integration is an achievable, or even desirable goal, when it clearly is not. Without probing deeply into the deep-seated motivations of young women like Hazel Massery and countless other white people and their visceral objection to integrated schools, we cannot appropriately address the segregation we see today.

At this stage of the twenty-first century, when we choose to acknowledge segregation, we attribute it to reasonable variables that deflect responsibility to intractable societal forces—housing patterns, class, historical inertia, zoning, parental preferences—everything except the deeply ingrained resistance of white communities to having their children educated alongside Black children. The absence of that direct historical confrontation allows us to quietly accept conditions that we will not publicly declare. It would be heresy at this time in America to advocate for racially segregated schools. But we have them. Having racially segregated schools is a prerequisite for advocating separate but equal schools—another heretical view in twenty-first-century America. Given that seventy years have passed since the *Brown* decision and schools remain racially segregated, the active resistance to integrated schools suggests that large segments of white communities still don't want them. I don't think segregation should be a goal. It is a concession to lesser American ideals. But I also don't think Black students ought to be banished to perennial suffering, waiting in unequal, segregated schools until one day when the glory comes.

"History repeats itself" is an overused cliché. It suggests that if societies do not learn from their pasts, they will repeat their mistakes. The associated overused cliché is attributed to Mark Twain, who supposedly said, "History doesn't repeat itself, but it often rhymes." The confluence of banning history and the surge of emphasis on AI in education leaves an opening for history to repeat itself. Because of the growing resistance within American right-wing conservatism and their control over state legislatures and school boards, there are now legal bans in modern America on the teaching of American history. Those bans prevent the educational interrogation and potential indictment of Hazel and what she stands for. They prevent us from learning from our past mistakes. They are blocking us from accurately reframing the problem of segregated public education because of an incomplete understanding of historical variables that are profoundly influencing the problem. The seismic retreat signaled by the overturn of the US Supreme Court's *Regents of the University of California v. Bakke* decision is evidence that the door is now constitutionally open for historical repetition. Related to segregation, one of the popular claims is that historical coverage of this era of American history is divisive and may make white children feel bad about themselves or their ancestors. It may run afoul of Georgia's HB 1084, the Protect Students First Act. More dangerously, there are popular claims that the segregation era was merely an aberration, a season out of step with American norms—that there was nothing structural or inherent about it.

Our failure to deeply assess the cultural causes and policies of racial segregation in schools has allowed us to use euphemisms to generate hallucinations of American public schools. The historic data is incomplete. As such, the problems, framing, and potential solutions are based on a hallucinogenic reality. The hallucination is softer and nicer-sounding. It is passive, as if no one is responsible for making it real. It is more readily receptive to unrealistic universal claims about what all students will be able to achieve based on their schooling. Most importantly, that false reality opens the door for history to repeat itself. In an era of AI, one potential outcome is the establishment of digital slavery and the creation of Black digital slaves.

It is unclear to what extent AI will transform the modern world as we know it. Future historians will have to look back and make that assessment. It is already true, however, that it has become one of the dominant topics in education. At first there were large district-level discussions on how to police students using AI-enabled tools. Plagiarism was top of mind as teachers and school leaders could not determine whether students were generating their work themselves or using tools like ChatGPT. The same uncertainty extended into foundational computing and mathematics courses. It was, and remains still, unclear how to determine whether students know basic coding or foundational mathematics, or are relying on AI-generated results. Here, too, is some historical repetition. The introduction of each new technical device that helped speed up and enhance our computational or cognitive outputs all raised the same questions—calculators, computers, internet search engines, etc. Now, school districts are contemplating "AI literacy."[16] The surge of emphasis on all things AI has the potential to usher in an era of Black digital slavery while we're not paying attention. Black digital slavery obviously isn't the same as the original American sin, but it rhymes.

What is Black digital slavery?

The captains of the modern tech world are not accidental leaders. They are armed with both vision and extremely advanced technical skills. Their ability to envision technical possibilities beyond the frontier is a combination of imagination, vision, and the technical expertise and courage to build things for which there is no precedent. In higher education circles today, there is a growing argument among innovators and futurists that the acquisition of technical skills is increasingly less dependent on formal postsecondary education. That is part of a much broader discussion about the future of higher education in the United States. The massive expansion of online degrees, courses, certificates, and credentials is evidence of this ongoing debate.[17] According to some surveys, as many as two-thirds of all higher ed institutions are adding online programs.[18] Of course, the COVID-19 pandemic was a driving factor, but in many corners, this expansion is driven by an equity argument. The ability to expand access to STEM-related skills to as many people as possible is an act rooted in equity and fairness. It democratizes the deployment and accessibility of technical skills. Another motivation behind online learning, albeit less publicly broadcast, is that it has the potential to be a significant revenue stream for institutions. But while these are big societal arguments, the narrower issue of acquiring advanced technical skills remains central to who really has access to the technological frontier and who does not—of who is creator, and who is slave to the creations.

In a crude reductionist framing of the issue, there are those who use technical tools and those who create them. In contemporary education debates around AI, it is the unspoken distinction between AI in education and education for AI. AI in education is generally the movement to find AI-enabled tools to facilitate the process of teaching and learning. It is the cornerstone of AI literacy. It will likely be the engine for the next wave of wealth generation in education as edtech companies vie to sell their wares to school districts and universities across the country—promising the world in the form of accelerated learning through smart, generative AI tutors; time-optimized teaching; individualized, student-driven learning; predictive, personalized advising; linguistic and auditory coaching; closure of achievement gaps; recovery from COVID learning losses;

reduced dependence on antiquated schools; mitigation of teacher and faculty shortages; facial recognition as replacement for in-school police; and pocket AI therapists to assuage mental health challenges and generate potential school shooter profiles. Ultimately, the promise is a shift in the modality of learning from institutions to individuals. It is a change in the unit of analysis. The claim will be, again, that technology is the great equalizer. The argument will essentially offer technical solutions to social problems under the banner of AI capabilities. Despite the august claims of the AI-enabled edtech wave that is surely coming, in the United States, the problems it is seeking to solve are rooted in the fundamental challenges of a broken education system, a system of segregated, unequal schools. For those students and teachers trapped in embattled, segregated schools, these AI tools will surely have some advantages and allure, but in the main, they are an opiate. AI in education, on its own, cannot provide students with an education for AI.

There are many ways to define what education for AI is. These are important curricular debates to determine the standards that will govern undergraduate and secondary school computing education in the near future. The complexity of that discussion and the variation of stakeholders render it both a critical national priority and a fraught one. Getting agreement across university faculty, across local and state districts, about what ought to be taught in an environment where the technology is changing so fast and constantly introducing normative questions about what students ought to know is an exceedingly difficult charge. Notwithstanding that, it stands to reason that the frontier of education *for* AI—the computing education necessary to build the tools themselves, to construct the algorithms and processes for combing through large language models, and to conduct cutting-edge research—happens at the nation's most elite institutions in computing. The top-tier higher ed institutions in the United States in computing are the institutions that are among the most elite and highly selective in the world. They're famous—MIT, Stanford, UC Berkeley, Carnegie Mellon, and the University of Washington, among others. And in some ways, they sit above the fray—inoculated from concerns about college-age students' demographic collapse or whether a college degree is still worth it, whether it still offers a return on investment. The American and global elite are still willing to spend fortunes, even cheat and risk jail, to have their children attend these US institutions of higher education.[19] Many of the best researchers and academicians on the entire planet reside at these schools. However elitist that may sound, as the saying suggests, it is what it is.

Leaning on the reductionist argument for clarity, these are the institutions where the frontier of education for AI and computational research is taking place. They are the institutions that have an intimate relationship with the intellectual engine of the Big Tech companies and the burgeoning AI tech sector. Sundar Pichai, the current CEO of Google, and its founders, Larry Page and Sergey Brin, all attended Stanford. Eric Schmidt, the previous CEO of Google, went to UC Berkeley. Sam Altman, the founder of OpenAI, the parent company of ChatGPT, attended Stanford. Kevin Systrom, the founder of Instagram, is also a Stanford alum, as is Jensen Huang, the CEO of Nvidia, and Reid Hoffman, the founder of LinkedIn. Bill Gates and Mark Zuckerberg, the founders of Microsoft and Facebook, both attended Harvard University. Jeff Bezos, the founder and executive chairman of Amazon, went to Princeton University. Evan Spiegel, founder and CEO of Snapchat, is also a Stanford alum. Jeff Dean, the head of Google AI, earned a PhD in computer science from the University of Washington. All of them studied engineering or computing or some combination thereof.

The companies they lead and the universities from which they come are, without question, the intellectual engines pushing the development of artificial intelligence and machine learning. These elite universities have a symbiotic relationship with the tech companies born of their alum and the start-up derivatives constantly launching from that intellectual orbit. Those companies are ever-shaping the computational landscape and pushing the mechanisms of modern human endeavor in ways that we are all experiencing. The transfer of intellectual capital in the form of research collaboration and the constant migration of faculty and resources back and forth between their campuses and companies, and the flow of graduates into the ranks of these tech titans, is its own special kind of intellectual inertia. It is a kind of technical slurry that may be unique in American history. In that sense, one form of education for AI is the sort born of this special mix of exceedingly high individual achievement at the nation's most elite institutions and their unique relationship with the most revolutionary technology companies in modern human history—and all of it wrapped in a mind-boggling concentration of wealth. It goes without saying that access to this special mix of intellectual and financial capital, computational pioneering, and cultural deification is exceedingly limited. It is one of the cloaked and gilded seats of contemporary global power—a Big House for the AI age.

It is worth repeating that, in 2024, two years after the release of ChatGPT, in the seven traditional public high schools that are essentially only for Black students in Atlanta Public Schools, on average four out of every hundred students were proficient in basic algebra. Said another way, more than ninety out of every hundred students in these schools do not adequately comprehend basic algebraic variable manipulation. The abstraction of objects in the form of variables and their subsequent relationship with each other is the cornerstone of computing and any STEM-related field. It is elemental to the probabilistic variable sequencing that is at the heart of generative AI. It is foundational to the formative mathematical maturity that is a prerequisite for existence as a technology creator. It isn't necessary to be a user. AI in education can serve these students, but education for AI cannot proceed without these basic fluencies. In two of the schools in APS where there was not a single white student, there was not a single Black student who was proficient in algebra.

If we take a sweep of the enclaves of urban segregation around the United States, we find similar patterns. In Detroit, one of America's most iconic Black cities, only 8 percent of primary school and middle school students are proficient in math.[20] Black eighth graders in Oakland, home of the Black Panther Party for Self-Defense, 12 percent.[21] Even in Boston, Massachusetts, in the shadow of MIT, 13 percent.[22] In Missouri, home to Ferguson and Michael Brown, may he rest in peace, 6 percent.[23] In the legendary cities of the Deep South, the anchors of the American Black Belt and the civil rights struggle—in Memphis, Tennessee, in Jackson, Mississippi, in Selma and Montgomery, Alabama—the stories are all versions of the same theme.

These data are analogous to a plantation fence. They rhyme with the anti-literacy laws of the slave era—the quiet, invisible theft of a foundational skill from thousands and thousands of young Black people. Segregation and the array of forces, policies, and people that enact it are trapping Black people in a world of digital slavery. They are blocking Black students from entry to the technical power enclaves as adults. There is causality in the relationship between segregation and level of academic performance, in this case in mathematics. Racial segregation *causes* academic gaps because of what racial segregation implies in contemporary America.[24] Of course, the issues are complicated and interconnected with concentrated poverty and teacher quality and school funding. Whole schools of education and research are dedicated to deconstructing the various causes of and solutions to racial academic achievement gaps. All the while, white supremacists might take this analogy to suggest that Black success is correlated to proximity to white people, and Black people on our own are simply not smart—a self-indulgent, delusional view. White mediocrity and underachievement are rampant in American society. But segregation remains a causal variable hidden in overly complicated speak. It is a plantation fence hiding behind high-minded specious vocabulary.

There are nomenclature wars—it is not an achievement gap; it is an opportunity gap. It is not a leaky pipeline; it is a siloed ecosystem. There are framework critiques—it ought not be referred to from a deficit mindset, rather, an opportunity construct. There are philosophical approaches—students need liberatory education, or abolitionist education, or anti-racist education, or citizenship education. Or they need inquiry-driven education, or problem-based learning. There are pedagogical approaches—they need integrated STEM education accessed through flipped classrooms and cross-class curricular continuity. They need balanced literacy as opposed to phonics. They need integrated mathematics as opposed to sequential mathematics, or spiral math versus mastery math. They need culturally responsive pedagogies. There are identity-based cultural affinity approaches—Black students learn more effectively from Black teachers. There are societal postulates—the policing of Black students in school reflects society's view of the innate criminality of Black children. There is the soft bigotry of low expectations. The teachers aren't teachers; they are facilitators in a community of learners. The students aren't students; they're scholars. Charter schools are the educational petri dish that will test educational innovation and spark transformation in the traditional system. Vouchers will give Black parents the choice and opportunity to send their children to school with white children, to get close—despite having no prescription for the illness of Hazel.

All of these approaches, axioms, and postulates swirl around the core American reality that public schools are segregated because, historically, white people did not, and do not, want Black people to live alongside them or Black children to learn alongside theirs. Tokens are fine, but real integration is not. There is a threshold of tolerance. When non-white residents begin to approach roughly 12 percent of a white community, that is too much. White people will leave.[25] And conversely, in the era of gentrification, when a neighborhood becomes sufficiently white, it signals that it is alright for other white families to come in and—to use crude reductionist language—take over. The smothering generational underdevelopment of Black schools by local and state jurisdictions saw to it that, despite our best efforts, high Black achievement, in most cases, was earned despite our schools, not because of them. Black students, in many ways, are simply trapped. It is the definitional foundation of segregation. It is the whole point. And in a season of AI, it is the foundation for Black digital slavery.

After all this, where are we? I return to Atlanta because it is the symbolic portal between the nation's civil rights past and its technical future—the home of both the King Center and Google data centers. As of the most recent census estimates in 2024, Atlanta's population is 47 percent Black and 40 percent white.[26] Yet, the public schools in Atlanta Public Schools are nearly completely segregated. We are right back where we started—separate and unequal. And in those few schools where both Black and white students attend, both segregation and diversity are only fleeting phenomena. The Black and white populations are merely passing each other by as the out-migration of Black Atlanta is making room for the arrival of the white landed gentry and suburban return-migrants from the era of white flight in the 1960s and '70s.[27] In the 2000 US census, the population of the city of Atlanta was 61 percent Black and 33 percent white.[28] Over the last two decades, the Black population has dropped 14 percent, and the white population has grown 7 percent.

There is no precedent in any major American urban center to suggest that this racial segregation and the out-migration of the Black population will not continue—choked out by a lack of the kind of education that makes for a viable future in tech-based economies. What future is there in Atlanta, for example, for the 88 percent of Black high school students who don't know algebra? What possible path is there to access the high seats of technical power that are creating AI systems, models, and tools? How can Black students overcome the inertia of segregation and penetrate the symbiotic relationship between the academic institutions that are developing and pushing the frontier for education for AI and the companies that are fueling and building off that intellectual capital? In the current mode, they cannot. Not in any meaningful representative numbers. Access to the elite institutions where the frontier of education for AI is taking place is essentially closed to Black students. Their populations at these institutions are evidence of that fact.

In an era when state superintendents of schools don't feel any compunction in banning African American history from being taught in schools, we are clearly again in an era when racist scorn for Black people does not need a cloak. There are surely people who, based on these data, would suggest that the common denominator in this problem is Black people themselves. They would likely argue as Herrnstein and Murray did in their widely discredited book *The Bell Curve*, and like so many racists before them, that Black people are inherently incapable.[29] They would likely argue that these are Black cities with Black leadership and Black teachers preparing Black students for Black jobs. They would overlook Hazel's lineage and the complicated and insidious components of a segregated system—and that segregation is merely the noun. To segregate is an active process of starvation and rebuke that renders these outcomes likely, if not inevitable. Now, given the constitutional overturn of affirmative action in higher education, there are many who will add to the noise of the season and suggest, "I told you so." They will argue that as soon as we stop considering race, Black students don't make the cut. They are less qualified, and in a meritocratic system, they don't deserve to be at these elite institutions. They will likely note the precipitous decline of Black students admitted at institutions like Georgia Tech, where there was a nearly 40 percent decline in Black admitted students in the first admissions cycle after the decision, or at MIT where Black enrollment decreased 10 percent in the same cycle. They would offer that as evidence that Black students are not qualified.

Lurking behind the alleged objective argument of blind meritocratic admission is the traditional American charge that Black people are simply not as smart. Of course, that refrain will likely change once it becomes clear that at these very institutions, white students cannot compete with Asian students. When white students become underrepresented minorities at the nation's elite technical colleges and universities, as is already beginning to happen, this argument about the supremacy of meritocracy will surely shift. However, at this stage of the American story, a story that is 250 years old, debunking the idea that Black students are not inherently inferior, but their schools are, isn't worth the time it takes to do it. The counterargument is simply racist, tantamount to nineteenth-century eugenics practices of measuring our skulls.[30] The reality is, as the common lament suggests, the system is doing what the system was designed to do. Intellect is not correlated with color. We all know that. But the sentiment remains hauntingly, quietly, and insidiously in the air. It serves as the paint on the plantation fence.

The confluence of forces that descend on Black students to render them isolated in segregated schools is a national tradition. It always has been. It is the sum total of generations of effort to make sure that Black students are isolated and educated for the lower tiers of American life. The exemplary cases of individual schools and students that have been remarkably successful do not define the truth; they mark the exceptions. Highlighting the 1 percent of heroic Black students who achieved the highest distinction in algebra in the Atlanta Public School system is as important and worthy of celebration as it is tragic. Hiding behind the outsized praise of those few students to claim progress made is what it appears to be—hiding. Right now, in an era when education for AI at the highest level is uniquely important, access for Black students is scarce, and AI in education—the mere use of AI-enabled tools—is abundant. Even if we resist dramatic hyperbole, we cannot escape the magnitude of what is happening. Who will build the next version of human intelligence?

If Hazel is building the DNA of AI, and Elizabeth is not, where does that leave us? It potentially leaves us with a history that doesn't repeat exactly, but with one that rhymes. It leaves Black people sitting idly, chaperoning Waymo's driverless cars, while the cars' systems relay data back to the tech creators to ultimately render the Black car sitters obsolete. It leaves Black people as Lyft and Uber drivers while the maps that monitor them learn how to optimize their routes and limit their pay. It leaves them riding mitted on scooters in the winter, using increasingly sophisticated apps to pick up groceries and deliver them to the Big House. Those are some of the outcomes of AI for education that we don't address in aspirational marketing. The noise and light surrounding AI in education and all it will do to close educational gaps are both deafening and blinding.

By now, at the 250th anniversary of the United States, we know enough to know that anything given freely, abundantly, and with enthusiasm to Black people ought to trigger suspicion and critical inquiry. What am I missing while I am distracted by these shiny new tools? Black digital slavery may be more difficult to see. Modern plantations won't have horses with white men astride them. They may well have the Constitution of the United States and the clamor of retrograde state legislators playing the role. Modern plantations may well have more deeply entrenched segregated school systems, with all the implications thereof, and efforts to both study and counteract those systems rendered illegal. Reading: illegal. Learning: illegal. History: illegal. Choice: illegal. Resistance: illegal. It may be a hallucination, but it all feels very familiar.

CHAPTER 2

DEMOCRATIZING COMPUTING CAME FIRST

At the end of the first quarter of the twenty-first century artificial intelligence reigns supreme in US STEM education. AI and its technical mystery and transformative power are quickly becoming the crown jewel of the STEM education ecosystem. The entire educational enterprise, from K–12 school districts to the most prestigious universities, and everything in between, is trying to carve a strategy for engagement with AI as it penetrates education. The enterprise is searching for ways to use AI-enabled tools to facilitate learning and determine what students ought to learn to participate in AI as an emerging STEM field in its own right. In some ways, in the popular imagination, the dividing line of the new dystopian world will be drawn between those who are AI-enabled versus those who are not. That idea is putting pressure on educational systems to ensure that their students are not proverbially caught on the wrong side of history. It is an age-old social pressure with each new technological leap forward. Those who adopt will reign over those who do not. Late adopters and non-adopters may be rendered obsolete as technology replaces their core skill set. History is littered with the casualties of slow adopters to technological paradigm shifts—Kodak didn't foresee computational optics, taxi companies folded under the weight of rideshare apps, entry-level software engineers are losing their jobs to AI tools that can already do what they do.

In the increasingly besieged community of advocates and educators concerned about Black students' participation in STEM fields, this new dystopian pressure is especially intense. Black students were already facing the precipice of irrelevance at the frontier of the tech enterprise. They already faced being a casualty of Big Tech paradigm shifts. Over the last twenty years, Black participation in core computing roles in Big Tech has increased slightly but remained abysmal. At Google, for example, the percentage of Black tech Googlers was 2 in 2014, and 4 in 2024.[1] At Facebook, Black tech workers made up 1 percent of the workforce in 2014 and 2 percent in 2021.[2] That community spent the better part of the last three decades building the movement for broadening participation in STEM fields as a cornerstone of civic and economic justice—an effort that, to date, remains incomplete. The movement was unfolding, however, in parallel with the changing social sentiment across the country that led to the rise of President Obama and the subsequent national racial reckoning as a result of the murder of George Floyd. There was a delicate consensus between the public and private sectors that broadening participation in STEM fields, and especially in computing, was a good thing to do. Racial diversity in the nation's educational and industrial technical enterprise was part of the social contract, albeit a reluctant clause. Those various conditions converged to yield a movement to democratize computing.

Now, the AI argument in education is being presented as a game changer—a miraculous technical solution to long-standing educational challenges. But it is being presented in a radically different American world—a world where the full-throated racist assault on Black people is back. AI, one of the most powerful technical advancements of our generation, is emerging just as the American pendular swing toward a more inclusive and tolerant society is reversing. The democratizing computing movement of the early 2000s, which focused on computer science and coding as essential twenty-first-century skills, has now given way to the era of AI in education. The fundamental problem, however, has remained unchanged. Without addressing structural inequities facing Black students in mathematics, for example, the promises of AI in education will likely face the same fate as other long-standing STEM initiatives. Notwithstanding the promise of AI in education, we risk creating another layer of technological exclusion while continuing to celebrate exceptional individuals as evidence of broader progress.

The dramatic transformation of the social context is important to the effort to bring justice to computing and STEM education. In 2016,

President Obama helped build the national case for computing as a new literacy. His argument centered on the fact that the economy was changing, and more Americans needed to be fluent in some level of computation for the nation to remain competitive. The name of the initiative launched by his administration was CS for All. It was the centerpiece of the democratizing computing education movement. The "for All" clause was considered and intentional. It meant that these uniquely valuable skills in the modern American economy needed to be accessible to all Americans. In building the broader case, he explicitly pointed out that Black people comprised less than 3 percent of the technical workforce in the United States at the time. And he offered the charge, "We can do better!" That was the voice of the nation's leader setting the tone and agenda for broadening participation in STEM fields, and in computing specifically, as a matter of national interest.

A decade later, during the presidential debates in 2024 between candidate Donald Trump and Vice President Kamala Harris, Trump declared on a national stage that Haitians in America, Black people, maraud through neighborhoods eating people's dogs. His very public assault on "wokeness" is an unveiled assault on Black people. This unfettered public racist animus toward Black people is a throwback to a different time in America. It is reminiscent of a time that, sadly, Black people remember intimately. It was a time when the government was the enemy. It reminds Black people of a time when white Southern politicians called efforts to dismantle Jim Crow laws the mongrelization of America, and elected government officials called on police to sic their dogs on us. It reminds us of a time when efforts to adhere to the *Brown v. Board of Education* decision led many white school district leaders to close their schools entirely rather than allow what they described as ravenous, salacious Black students near their virtuous white children. It reminds us of a time when national politicians referred to us as super-predators. It is in this national context, when the nightmares and ghosts of America's recent past are on the loose, that we enter the AI era in education. The significance is that there is no national appetite in this season to ensure that the "for All" clause in fact means *all*. On the contrary, the current national tenor coming from the highest levels of the American government offers that the sum total of Black people's accomplishments is a gift from charitable white people who offered us opportunities we are not qualified for simply because we are Black. As such, we should be relegated to and grateful for what President Trump refers to as "Black jobs."[3] In the current federal framing, those do not include seats around the creative desks of the most powerful technologies of our generation. If we are granted subpar education, it is in keeping with a federal expectation of our subservient role in American life.

There is tectonic pressure here. In the technology world, there is a cultural instinct to claim that technology is the gilded sphere of merit and objectivity. In the most reductionist view in the tech space, something either works or it does not—in its foundational mathematics, the numbers either add up, or they don't. Social and cultural matters allegedly reside outside this rarefied space. It is purportedly the temple of meritocracy. In education circles, there is a tendency to speak of mathematics and engineering education in isolation, as if they have a special inoculation from the racial patterns and cultural forces that shape the correlation with race and nearly everything in American life. Ironically, AI brings with it a host of potential dangers that sit at the intersection of technology and persistent social and cultural ills. Its advance into education is happening at a time when state legislatures are attempting to remove and revise the core role of slavery and race as the foundational epicenter of the American story. The claims of advocates for AI in education—that it will offer remarkable pedagogical advancement and curricular enhancement—must be measured against these larger structural variables. The application of technical solutions to social problems has potential, but not when the nature of the social problem is underdefined, or worse, erased. In that sense, AI in education is a continuation of the attempt to democratize computing. The unfinished effort to democratize computing offers important context to the challenges ahead with AI in education.

The effort to democratize computing plays a unique role in the most recent phase of US social and educational movements because of the rise of Big Tech. Because of the rise of Big Tech, democratizing computing has been a top priority in STEM education for the past twenty years. It is not unique, however, as it sits in a long history of efforts to close achievement gaps in various disciplines in American education for the purpose of broadening participation in the ever-evolving workforce. Reading, writing, and arithmetic are the cornerstones of basic education, and there are long-standing efforts to improve American education, particularly public education. In education circles, the disciplinary battles and haranguing over how best to teach mathematics or literacy or how to integrate science or problem-based learning across curricula are never-ending. All the while, the long arms of racial segregation and racial animus toward Black students in American schools continue to strangle any systemic efforts to broadly make the education of Black children equal to that of white children. In public education in America today, there is no nationwide or state-wide data on any subject, on any metric, at any level where the average achievements of Black students and white and Asian students are indistinguishable. Notwithstanding the tortured handwringing of school leaders to parse and control for this or that variable, or speak of gains instead of mastery, they know, as do the parents of children in schools all across the country, this is an inescapable American stain. This basic divide is acutely true at the highest levels of academic achievement. None of the portals to the most elite rungs of modern education—gifted programs, honors courses, Advanced Placement courses, International Baccalaureate Diploma Programmes, elite high school entrance exams—are exempt from the long arms of race.[4] The racial differences in participation and achievement at those levels are soul-crushing. I'll return to this point in chapter 3. In the main, Black student achievement at the highest levels continues to be despite the education system, not because of it. This unfortunate reality has given rise to concepts of liberatory education, revolutionary education, citizenship education, culturally responsive pedagogies, hip-hop, fugitive and colonial pedagogies, and a host of academic attempts to reframe and deconstruct a system that was designed at the outset to either not serve Black students at all or undereducate us. With this history, efforts to close racial achievement gaps are nothing new. They are a fixture of American education because the foundational structure of the system remains intact—segregated and unequal. The system is doing what it was designed to do.

Given that context, democratizing computing education is yet another attempt to close gaps, open access, and broaden participation in a particular area of American education—computing and computing-related education. It is distinguished in this century because of the shape-shifting influence of Big Tech—of Google, Apple, and Amazon, for example—that has catapulted computing to the epicenter of valuable skill sets. Democratizing computing means making computing education broadly accessible.

Factors such as race, class, and geography ought not be deterministic variables in whether students have access to high-quality computing education. At this stage of the American story, however, there is hardly a need for more evidence to demonstrate the correlation between race, class, and geography and every indicator of life. In that sense, the particular effort to find a path for Black students to participate in high-end computing faces the challenges that all efforts to democratize anything in education face. It is an attempt to close gaps of access, achievement, and expectation that are built into the system that produces the gaps in the first place. That is not an observation of fatalism but a grounding acknowledgment of the social dimensions of the challenge. Like other educational efforts, the attempts to close gaps and open access occur through some combination of policy and programs. They are a mix of systemic variables and local ones. It is the range between having computing as a high school graduation requirement in a state and offering a summer coding camp for kids. There is not much new about this reformation formula or its modern limitations. The effort, however, still takes place in a social context. In today's America, the president of the United States asserts that Black people eat dogs, diversity and equity are criminally divisive, and Critical Race Theory is the ideology of radicalized terrorists. This is the social context in which the shift from democratizing computing to AI in education is taking place.

I am on the faculty at Georgia Tech. I also worked at Google for several years as the director of STEM education strategy. Prior to that, I was a program officer at the National Science Foundation (NSF) in the Directorate for Computer and Information Science and Engineering. While there, through the Office of Science and Technology Policy, I served on President Obama's My Brother's Keeper Task Force for STEM + Entrepreneurship. Before that, I was the director of the STEM Initiative for the University System of Georgia. These are all institutions that play an outsized role in American STEM education, its utility, and the engine of the modern technology economy.

Georgia Tech is one of the premier technical institutes in the United States and is the most prestigious in the Black Belt of the American South. It was founded in 1885, twenty years after the end of the Civil War. It sits physically in Atlanta, which is marked as much by memorials of civil rights giants John Lewis, Martin Luther King Jr., Ralph Abernathy, and Joseph Lowery as it is by Stone Mountain, a sacred site of the Ku Klux Klan and the largest stone relief carving in the world. The carving memorializes Robert E. Lee, Stonewall Jackson, and Jefferson Davis—captains of the Confederacy and defenders of American slavery.

Georgia Tech is the flagship public higher education institution in Georgia. That is significant because, even now, at the end of the first quarter of the twenty-first century, 33 percent of the population of the state is Black, a vestige of its location in the American South. The Black population in California, by contrast—arguably the epicenter of the AI movement in North America—is 5 percent of the total. Despite these population proportions, the student population of Georgia Tech remains around 8 percent Black and has hovered around that level for the better part of the last twenty years. In fall 2025, 7 percent of the students in the College of Computing are Black. In sum, Georgia Tech personifies the challenge. The institute sits at the frontier of the AI race in research and education but has its roots in the American South, where racial hierarchy and terror were founding principles, and the largest memorial in the entire state is for Confederate heroes. The confluence of these large patterns of racial participation and cultural context helps bring the big lofty ideas of democratizing computing, AI in education, and STEM education down to earth.

I operate in a world where STEM education is the organizing principle and lifeforce of what we do. I was in the Speed Track to Engineering Program at Midwood High School in Brooklyn in the late 1980s. I earned bachelor of science and master of science degrees in mechanical engineering and then got a PhD in science and technology policy. My entire education and professional career have been built on and centered around STEM education in some form. But I do not assume that the centrality of STEM education in the world, as I see it, is universal. It is not. Nor should it be. I am not making this about the significance of STEM education simply because I think it is important, though I do. It is because STEM has become a cultural force in the first quarter of this century. The importance of STEM education, STEM jobs, STEM careers, and even STEM identities over the last twenty-five years has been ringing out from all educational quarters, including economic development planners, summer camp marketing materials, school certifiers, academic advisors, parents, and every hill and molehill between Silicon Valley, California, and Research Triangle Park, North Carolina. While AI in education may be new, the whole endeavor is merely the next iteration of a cultural wave. That wave is the ascendency of the importance of STEM education and all things tech-related. It has historic roots in national security, but it has become a kind of feature of middle- and upper-middle-class life. The student who says to their parents' friends that they study computer science at UC Berkeley receives a pause of awe and adoration, while the one studying English receives a pause of pity and perplexity. This AI phase is unique in how deeply it has penetrated the psyche of Americans, but it is building on a long-standing effort to democratize computing and expand access to STEM education to diversify and bring more Black students into the STEM enterprise. Here, too, history matters. Some assessment of the state of affairs, and some brief look back, is warranted as we leap forward into this new season of AI in STEM education.

In the beginning, the popular push for improved national STEM education was driven by concerns about the Cold War and competition with Soviet science beginning in the 1950s. Subsequently, as the Cold War cooled, global technical innovation and dominance became the national priority for economic and national security reasons. Interest in computing education as a matter of national security is steadily increasing because international cyberwarfare and the sophistication of AI-enabled weapons are shifting the skills required for modern warcraft. While federal prioritization of STEM education in the workforce is the wonky stuff of the Office of Science and Technology Policy and the United States Digital Service, the popular cultural penetration was partially ignited by the spread of global fear of the allegedly world-ending Y2K bug.[5] As 1999 wound down and we approached the end of the twentieth century, the news filled us with fears of total catastrophic computing and software collapse at the stroke of 01.01.00. The general population was beginning to appreciate the ubiquity of computers and computer-driven systems in our lives. We were beginning to see how dependent we were becoming on computational technologies in our everyday lives. It is hard to remember that, in 1999, Google was only a year old, most people had not heard of it yet, and it certainly was not a verb. Facebook wouldn't be founded until 2004, and the first Apple iPhone wouldn't hit the world until 2007, eight years later. Neither phones, cars, houses, nor cities were "smart" yet. IBM and Microsoft were the dominant technical titans at the time. STEM education as part of broad American educational culture was still in its infancy.

In those days STEM education focused heavily on engineering. Computer science as a field had not yet taken on the social supremacy it now holds. As the internet began to grow and the penetration of technology into our lives began to deepen, the cultural relationship with STEM education began to form. The dominance of Microsoft and the expansion of Google and its new tech peers began to be popularly associated with the smartest among us, financial success, and the new American working rich. Because of that, and what appeared to be the magic of the internet, the emphasis on STEM education began to surge. It was beginning to be seen as a path to the top tier of American life—finally a competitor to the all-out money worship of finance. The mantra at the time in education circles was "We need to prepare students today for the twenty-first-century STEM jobs of tomorrow." That declaration from parents in town halls and in meeting after meeting with school leaders and philanthropists was accompanied by the clause, "and technology is changing so fast, the jobs of tomorrow don't exist today." It was said that we in the education community needed to be able to provide students with the kind of STEM education that would render them lifelong learners so that they could be continually relevant in a swiftly changing technical world.

While that cultural shift was taking place, the issue of diversity was ever-present. There wasn't yet the array of diversity positions that there are now. Diversity at the time was merely an uncharged noun. Vice presidents of diversity and chief diversity officers emerged later in the 2000s and exploded in the post–George Floyd era. In the early days, the diversity efforts in engineering were riding the inertia of the Civil Rights Movement. Bob Moses, an American hero of civil rights, voting rights, and mathematics education, in his 2001 book *Radical Equations* warned that Black people were being offered a sharecropper's education and being kept there, in part, by a lack of proficiency in basic mathematics.[6] The expectation of public education, he argued, was that Black people would operate in subservient roles, and our education was commensurate with those expectations. Jonathan Kozol's *Savage Inequalities* from a decade earlier in 1991 made a similar case.[7] Kozol, one of the preeminent education writers of the second half of the twentieth century, laid out in graphic detail the horrible conditions Black students faced in public schools. Black students' ability to climb out of such educational misery was a matter of justice, civil rights, and the righting of American wrongs. Addressing equitable outcomes and expanded participation in STEM fields required coming to terms with the kind of education most Black students were receiving—which was terrible. Here, too, there are tomes of research about the abuse that is the systematic undereducation of Black people—from Carter G. Woodson's *The Mis-Education of the Negro* to Jarvis Givens's *Fugitive Pedagogy*.[8] Woodson and Givens, both education scholars separated by nearly a century, make the same case—not only are Black students subject to deplorable schools, but what they're being taught in those schools amputates their ambition and crushes their sense of pride and value as Black people and as American citizens. At this point in the American story, that kind of research remains important evidence, but it merely affirms what many people, especially parents of Black school-aged children, already know. For distinctly American historical racial reasons, Black education in America is a complicated mix of something to be freed from and something to be dependent on. It holds both our chains and our hopes.

While the early advocates for diversity in engineering and STEM fields toggled between their north star arguments—whether diversity was good for business or was an act of justice—they were clear about the critical role of mathematics. They were unequivocal in their assertion that mathematics was foundational to any meaningful participation in postsecondary STEM education. In the early 2000s, STEM was not yet a single word with an ambiguous meaning. People didn't yet meaninglessly declare that they were going to "study STEM" or "do STEM." It was still an acronym for science, technology, engineering, and mathematics. Pulling apart the acronym makes it easier to see. Engineering is not the same thing as mathematics. You cannot earn a bachelor of science in STEM. The fields are distinct, with different educational needs and certainly different social dynamics. Gender challenges in biology, for example, are not at all the same as those faced by computer science. There is also disciplinary hierarchy. None of the science, technology, or engineering disciplines are possible without foundational mathematics. You cannot study mechanical engineering without proficiency in calculus. But you can study it without knowing anything about biology. In those days, the argument was more simple and honest. Mathematics is the natural root language of all STEM fields. In order to diversify the STEM enterprise structurally, we needed to address the proficiency of Black students in foundational mathematics.

Upon completion of a master's degree from UC Berkeley, and before I continued at Georgia Tech, I worked at the National Action Council for Minorities in Engineering (NACME) from 1995 to 1997. It was one of the early entrants into the world of advocacy for diversity in engineering and STEM fields. Founded in 1974, it was one of a groundswell of organizations born in the post–civil rights era to serve my generation, the first generation of legally free Black people in the United States. They sought to enlighten Black students on what engineering is, what the educational requirements are, and carve the paths to get us there at the highest levels. The National Society of Black Engineers (NSBE), the largest student-led organization in the United States, was founded in 1975. The Black Engineering and Science Students Association (BESSA) at UC Berkeley started in 1968—the same year Martin Luther King Jr. was assassinated. INROADS, an organization designed to get more Black students involved in the American corporate structure, and with an emphasis on technical fields, was founded in 1970. The AT&T Bell Labs Fellowships started in 1972. The array of organizations and corporate-sponsored internships and scholarships at the time defined the experiences of a generation of Black students studying engineering. We, of that first generation, arrived in college in the late 1980s and early 1990s. Our parents were immigrants or went to legally segregated schools. We emerged into a new world. What system did the nation have in place to accommodate its newly minted ideal that Black people were full citizens and were entitled to the full spectrum of the American educational enterprise? That was the context of the early efforts to diversify the STEM educational enterprise.

As is the case in any broad-based educational effort like this, there is a swirl of interconnected variables. All of these organizations sought to address the complex sieve that Black people were slipping through, preventing us from getting, in large numbers, to the frontier of American technical progress and prowess. Notwithstanding the complicated role of race, which complicates everything in American life, mathematics was foundational. Without proficiency in mathematics, none of these efforts could be broadly successful. It is an impossible task for any student to be successful in any STEM field without basic mastery of mathematics. It is debatable what level—whether calculus, statistics, or algebra—but mathematics remains foundational. Successful efforts to broaden participation in STEM fields require successful efforts to improve students' achievement in mathematics. Beyond Black people, mathematics proficiency is a central pillar of the risk that the country now faces in its technical competition with China. White American students, long the default image of what an American student is, are underperforming in mathematics, too, when compared to their Asian contemporaries. In 2023, white students scored, on average, nearly one hundred points less than Asian students on the math section of the SAT.[9] But the problem for Black students is extreme.

In 1996, in conjunction with the Ad Council, NACME launched the Math is Power ad campaign. It was designed to popularize the importance of algebra and help students make the decision to opt in at their schools. It was targeted at students themselves in a grassroots approach. It was based on the simple assertion that any success in getting Black students to participate in postsecondary engineering is constrained by their access to and success in mathematics—namely algebra. Math is Power was a loose reference to the Black Power slogan of the 1960s. Stokely Carmichael's mantra, combined with the raised Black fist of the Black Panther Party for Self-Defense and their chant—"Power to the People"—was the thinly veiled subtext for the ad. It was a picture of a fist with the letters "M-A-T-H" emblazoned on the knuckles of each finger.

Also in 1996, one of the very first assignments I was given at NACME was to talk to a group of high school juniors and seniors in the Bronx about engineering and make them aware of the programmatic support NACME could offer them. I was helping to run a program NACME offered at the time called the Vanguard Program. It was designed essentially to have me and a colleague travel to high schools in several major cities, especially New York and Houston, to find Black and Latino students who wanted to pursue engineering—and offer them comprehensive financial and academic support to do it. In most cases, we had to explain what engineering was first. The criteria for selection to the program were based on how students worked in groups to solve an array of math problems. Once selected, a major component of the summer residential program—as was typical in those days—was an intensive calculus immersion. The students needed to be fully aware of the depth and pace of collegiate-level calculus. One element of the logic model of the program was that calculus was determinative of students' success in any of the engineering fields they chose to pursue.

This experience was deeply instructive. I was hoodwinked at this first step into NACME life. When I was sent to this Bronx high school, I was under the impression that I would be speaking with a small group of juniors and seniors about engineering and college. When I got there, the entire senior class, with some invited juniors, was jammed into the auditorium. I was introduced by the principal, handed a mic, and sent to the middle of the stage in front of three or four hundred high school students for the whole assembly period. Even though this was nearly thirty years ago, and I don't remember the name of the school, I remember the principal saying that I had done engineering at UC Berkeley and "he is just like us." That struck me, and I remember it. How did he know? What was it about me that made me just like the students in this school? No, I wasn't. I'm from Brooklyn. Not the Bronx. Notwithstanding that egregious mistake, it was because I was Black. That was all. That's all he knew about me.

It was my early exposure to the overly simplistic racial transference argument—if he can do it, so can you. Students "need to see it to be it," is how the saying goes. There is merit to that argument for sure. It sets the foundation for what is normal in the quiet chambers of a student's imagination. It is necessary to break the generational chains on our imagination. When we close our eyes and envision a person in handcuffs with their face pressed against the wall, their legs kicked, spread-eagled with an NYPD forearm on the back of their neck, do we see a thin white woman named Amy in yoga pants? No, we do not. That image is so far-fetched it is almost comical. Or do we see Yusef Salaam of the Central Park Five, who looked like I did as a Black teenager? When we close our eyes and envision an engineer, does an image even come to mind? Do we see a Black man with locks, as I was? Or do we see a white man who looks like Bill Gates? That kind of primitive private envisioning and image disruption is important. It informs what we expect of ourselves and what we see as possible. In that sense, the principal was right. I was just like they were. However, that popular argument often overlooks the several structural barriers that determine success.

Conversely, it sets our expectations of what ought not happen. In Nelson Mandela's autobiography, *Long Walk to Freedom*, he describes some of the insidious outcomes of this kind of image reinforcement.[10] He described how Apartheid played tricks on Black people's expectations for themselves. So accustomed had Black people become to their own suffering and pain and the images and experiences of it, they'd become numb to seeing it. They could see suffering, and despite every effort to stay out of it, they could slip into it themselves, and it would be sad, but it would be normal. He goes on to describe an experience of seeing a wretched white woman on the street. She was poor and disheveled. He described how she triggered a surge of pity in him. He recognized that, while his cause was Black freedom, he'd become so accustomed to seeing Black suffering that it didn't trigger individual pity. However, seeing a white woman suffering was so out of the ordinary, so misaligned with societal norms, it felt as if something was broken in the system. Having Black people as the face of the Vanguard Program doing the hard-core mathematics and articulating the contours of the engineering world was certainly important and intentional. It was an attempt to break something in the system. The images associated with Black people, especially in the 1990s as the country was at the height of the crack epidemic, were laced with criminality and underachievement. Images of Black engineers from Brooklyn, having attended one of the premier engineering schools in the country, were a glitch in the system. They were designed to disrupt the standard imagery and correlation with race and technical accomplishment.

While the racial transference argument is important, this experience in the Bronx high school set the limits of the argument for me. Yes, it is important to know that Black people have achieved at the highest levels in all fields and have done so despite tremendous obstacles. That Vanguard experience was twenty years before Barack Obama rose to the United States presidency. The ascension of President Obama flooded us with the racial transference aspirational story. A Black man is the president of the United States, a Black man holds the most powerful job on the planet, a Black man is the head of the free world—ergo, Black man, Black woman, there is no limit to what you can achieve. We were flooded with the spirit of Marcus Garvey's mantra, "Up you mighty race, and accomplish what you will." The trouble with that story, with the principal telling his students I was just like them, is that it is too simple. He's Black, you're Black, we're all Black, we can all be engineers. It bestows a misplaced significance on race. Race is absolutely a deterministic variable in American life. The country has racist demons that are alive and well. But used in this way, it is a distraction from the more insidious consequences of race.

I may or may not have been an inspiration to the students in that school, but as my discussion with them and their teachers unfolded, I learned that the school didn't offer calculus, and virtually none of the hundreds of students in the school had passed algebra. While yes, I am Black, I also took Advanced Placement Calculus BC in the eleventh grade and differential equations in the twelfth. The greater significance of their being Black was not that they might somehow have a racial or aspirational affinity with me as an individual; it was that they were allowed to be in a school that didn't offer calculus—a foundational course that is a prerequisite to virtually any postsecondary STEM field. At that time at UC Berkeley, students without calculus were ineligible for acceptance into the College of Engineering. What I was bearing witness to was the execution of a crime. The notion that I am Black, that they are Black, and that, therefore, our respective ambitions were mutually reinforcing was a sham. If they did not have the appropriate level of mathematical achievement and maturity, what I was offering was false hope. Of course, there would be exceptional students who, against all odds, would succeed. They would succeed, however, despite their school, not because of it. They would be the charmed and favored Black students who appear on the emotionally charged brochures of scholarship programs across the country. Here, but for our scholarship aid and commitment, is at-risk, underrepresented, minority, African American Yusef. He is an engineering major at said university, or an engineering intern at said corporation. He is exceptional. The fact that he was rescued from a crime scene is lost in that story. The Black affinity argument is emotional, and I respond to emotional inspiration in very visceral ways. But I am not fooled by it. The challenge is fundamentally structural and reinforced by cultural expectations. It is a social problem.

It turns out that nearly thirty years later, I would serve on the Board of Trustees of Spelman College in Atlanta. Founded in 1881, it is one of the most prestigious colleges in the United States. Its legacy is a continuous provision of education for Black women, beginning in Reconstruction, coming right out of slavery, until today. One day on campus, I just happened to see on the wall of one of its storied halls the NACME poster. The very poster from the Ad Council in 1996—Math is Power. Math was, and still is, power. It brought me right back to the school in the Bronx and the limitation of the racial transference argument. I saw again the difference between emotional and inconsequential exceptionalism and the provision of real substantive education. On the poster, beneath the fist with "M-A-T-H" emblazed on the knuckles, was a phone number. If you were, as a student, or as a school administrator, interested in the Math is Power campaign, you could give NACME a call. I called the number in 2023—twenty-seven years after the ad campaign. It rang. It was to a car dealership. Again, in 2023, in Atlanta, where Spelman College sits, in the seven schools in the Atlanta Public School system that are made up of nearly 100 percent Black students, on average only 4 percent of them are proficient in basic algebra. The phrase "If you can see it, you can be it," to a pioneering people, may or may not be necessary, but it is certainly insufficient.

NACME still exists and is doing the ongoing work of providing Black and Latino students paths into postsecondary STEM education. In some ways, the battle must be ongoing. We cannot wait for a magical reformation of public education in the United States that will equally and justly serve Black students. That is a fool's timetable. There are students in schools today who require our advocacy and championing and inspiration in whatever form we can offer. If seeing an accomplished Black person ignites the spark in some student, even one, it is worth it. The cliché analogy is of the person saving crabs from swarming gulls. It may not make a difference to them all, but to the individual saved, it meant the world. The larger point, however, is that the fundamental challenge that all those early advocacy groups identified—the challenge of Black achievement in mathematics—has remained unsolved. The path to postsecondary STEM education, especially at the highest levels, is an extremely narrow one for those students without fluency and competency in foundational mathematics. That is a cold, hard truth.

Despite not having resolved that basic problem, the part of the STEM education ecosystem that has a focus on justice is marching on. Its goals have morphed with the times. The ascension of Big Tech in the twenty-first century and the rise of computing have simply shifted the goals of many of the organizations from engineering and STEM fields broadly to computing specifically. The goal now is to carve a path for

Black students to find entry into the world of computing-related disciplines. "Code is Power" became a contemporary mantra. Coding came to be viewed widely among educators, education entrepreneurs, and philanthropists as the essential skill of the twenty-first century. We might envision the same math fist with "C-O-D-E" emblazoned on the knuckles instead. In the late 2000s and in the following decade, the STEM education community went through a transformation. Coding bootcamps, coding summer camps, coding certification programs, coding academies, and coding advocacy groups were ascending. Every version of coding was available. Black Girls Code, Girls Who Code, YesWeCode, Code2040, Code.org, and All Star Code were all advocates of the same thing—trying to carve a path to the world of computing for students to gain access to the ascendant tech economy.

The first twenty years of the 2000s were the era of democratizing computing education in STEM education circles. Within the community of computer science K–12 teachers and university faculty, there are qualms about the nomenclature. What is the difference between coding, computer science, and computing? Within their academic conclaves, those differences are very real. While to the general public those differences are meaningless, there are legitimate questions of scope and sequence, content and pedagogy, that distinguish them. For degree-granting institutions, you cannot earn a degree in coding for example. There is a rebuke of the reductionist treatment of computer science as coding that suggests that anyone can learn how to code in essentially a few weeks—an argument promoted by the marketing arms of several coding bootcamps and corporate certification models that sprang up during that period. It feels demeaning to the complexity of computer science. And all of it, of course, is now threatened by AI-enabled tools. Notwithstanding those important questions of syllabi and curricula and degree criteria, in the public domain, the focus was on coding as the holy grail. It was the concept that was marketable and easy to describe and served as the bedrock of the equity-in-computing focus for the next phase of the STEM education movement. Students needed to be able to learn how to code to be able to participate in the tech economy. That was the basic logic of the movement. For advocates of equity-in-computing fields, the argument had the standard racial focus. The surge of video games and the innumerable apps that were coincident with the ubiquity of smartphones provided for the distinction between a technology consumer and a technology creator. Advocates declared that Black students and communities ought not be merely consumers of technology, subject to the whims and dangers of uninformed consumption. They ought to be involved in the creation of these technologies, to exert some influence on the future of the industry and the construction of its regulatory guardrails. But perhaps more importantly, to be involved in the shaping of the modern American economy and technological landscape of power and wealth.

This was an important stage in the development of computing as a cornerstone of modern STEM education. Legend has it that in 2014, President Obama was the first sitting president of the United States to program a computer while in office. In 2016, in his State of the Union Address, he stated that "offering every student hands-on computer science and math classes that make them job-ready on day one" is a national priority.[11] It was a recognition that the epicenter of power in the global economy was shifting from natural resource extraction and banking to information commerce and technology. He was directing the nation's education system to prepare itself and be responsive to global economic shifts.

In 1990, the five most valuable companies in the world were all Japanese, and four of the five were banks: Nippon Telegraph and Telephone (NTT), valued at $150 billion; Industrial Bank of Japan, valued at $104 billion; Sumitomo Bank, valued at $73 billion; Fuji Bank, valued at $69 billion; and Dai-Ichi Kangyo Bank, valued at $64 billion. Ten years later, at the turn of the century, the shift to the United States as the epicenter not only of the global economy but of technology-based industry was well underway. In 2000, the top five most valuable companies in the world were all American companies; none of them were banks and three of them were explicitly based on high-end technology: Microsoft, valued at $587 billion; General Electric, valued at $460 billion; Cisco Systems, valued at $455 billion; Exxon Mobil, valued at $302 billion; and Walmart, valued at $275 billion.

The early 2000s marked the birth of Big Tech, and eight years later President Obama foresaw that the nation's workforce had to be compatible with the comprehensive transformation to tech dependence in nearly all industrial sectors. He instructed the President's Council of Advisors on Science and Technology (PCAST) to produce a road map to ensure that the United States maintains technological leadership in the changing global economy. They produced the report *Engage to Excel*, which outlined a series of nationwide education recommendations to achieve the president's goals.[12] He saw the issue not only as a matter of continued global leadership of US industry but as a means of limiting the yawning chasm in the American workforce between those capable of participating and those excluded from the technological transformation. That chasm fuels social tensions and animus when the gap between the rich and the poor becomes extreme. Historically, that gap pits populations who feel entitled to a good life against those they think are stealing that life from them. In the American story, that is white communities intensifying their racial vitriol against Black people and other communities of color. President Obama viewed this emphasis on STEM education not only as a necessary economic workforce input but as a means of contributing to social cohesion and tolerance by keeping the gap between the working rich and the working poor as small as possible.

In 2020, the five most valuable companies in the world were Apple, valued at $2.1 trillion; Microsoft, valued at $1.6 trillion; Amazon, valued at $1.6 trillion; Alphabet (Google), valued at $1.2 trillion; and Meta (Facebook), valued at $800 billion. By 2020, the transformation was complete. All of the top five most valuable companies in the world were American Big Tech companies. The engines behind them all are fueled by a workforce with advanced computing skills. The tech sector had become the leading American and global enterprise and was rapidly emerging as the epicenter of fantastic wealth and power—the magnitude of which rivaled that of nation-states. If Apple were a country in 2020, its value would have positioned it as having the sixth-largest GDP in the world, ahead of Italy, Brazil, and Canada. Broader participation in this emerging gilded sphere was the goal of democratizing computing education. Of course, there are legitimate questions about whether the accumulation of such fantastic wealth and power among such a small group of people is healthy for a democracy at all. Notwithstanding that, leaving the social structure in place and those questions unanswered, it was increasingly clear that if Black people were going to participate in this new power sphere, democratizing computing was a contribution to that end. President Obama was perhaps unique in his explicit recognition that Black students faced very specific structural and systematic challenges, as well as that the significance of computing as a path to modern power was inarguable.

In 2015, he signed the STEM Education Act (HR 1020) into law.[13] The act was a federal recognition of the changes occurring in the STEM education world—that computing was beginning to dominate the industrial needs of the nation and the imagination of educators across the country. The act sought to explicitly include computer science as one of the "STEM" fields. The significance was that when federal efforts to bolster STEM education were targeted at various federal agencies in the interest of national priorities, computer science was expressly included in that. Computer science had become so important that the president passed a law to make it explicit. It outlines the definition of STEM education:

> For purposes of carrying out STEM education activities at the National Science Foundation, the Department of Energy, the National Aeronautics and Space Administration, the National Oceanic and Atmospheric Administration, the National Institute of Standards and Technology, and the Environmental Protection Agency, the term "STEM education" means education in the subjects of science, technology, engineering, and mathematics, including computer science.

These efforts at the federal level were likely unseen and unknown to the general public, but they represented a national directive to democratize computing. It was a signal from the highest level of the federal government that computer science is of growing importance and will be transformative for the future of the United States in this century. Until that point, computer science education in public K–12 schools was mostly the purview of highly talented students in highly privileged schools. It was not regarded as a fundamental literacy. Given the state of public education in the country and its ongoing struggles to effectively educate American children even in the basics of mathematics and literacy, computer science was, and continues to be, a stretch. It is a delicate educational directive because it assumes so many other fundamental skills are in place. Also, in nearly all states at the time, there were no accountability measures. Neither schools nor teachers were held to account for the quality or breadth of CS offerings.

American first graders who started public school in the fall of 2002 were the first to begin under President George Bush's No Child Left Behind Act (NCLB). They graduated high school in 2014 as the first generation of American public schoolers for whom their entire K–12 education occurred under NCLB's relentless emphasis on standardized testing and the narrowing of education to focus on accountability. That era of public education in the country, as all parents of children in public schools now know, placed an extraordinary and punitive emphasis on standardized testing. Schools' fates and their ratings, and the jobs of principals and teachers, all hung in the balance of standardized testing outcomes. At the local level, if schools did not meet their Adequate Yearly Progress (AYP) goals, they could be closed, have all staff fired, and be taken over by the state—as if the state had some magic wand that they'd chosen not to wave previously. Of course, the violence of this punitive warfare was taking place in the nation's high-poverty enclaves in nearly all major urban centers from coast to coast. Those were the schools where concentrations of Black students trapped in poor communities struggled most with accessing effective, high-quality education and achieving the AYP goals. Like so many other things, the NCLB wars and the stakes of the punitive measures were most consequential in the most vulnerable communities. It was a fraught era that gave rise to cheating scandals, the charter school movement, and Republican efforts to promote school choice and siphon public funds away from public school systems to private and parochial schools.

The punitive and political threats put extraordinary moral and ethical pressure on teachers weighing the solvency of their careers against unrealistic expectations of miraculous student progress in generationally underperforming schools. The education of Title I students, the children of America's poor and beleaguered, was under intense scrutiny, and the arguments for and against NCLB raged in academic education circles. The basic premise of accountability many parents and educators agreed with. President Bush's indictment of the "soft bigotry of low expectations" was real and appeared heartfelt. The reality, however, was that it placed blinders on public education. Anything that did not lead to higher performance on standardized tests was not an educational priority. It muted the colors and contours of a complete education for American students. They were locked into a perennial rhythm of year-long test prep from September to April, then exhale until the end of the year and pray the AYP results were sufficient. President Obama's push to include computer science in the foundation of American public education not only had to confront the popular view that computer science was the purview of "nerds" and sun-deprived, quirky white men but also the juggernaut of the NCLB accountability system, which prioritized certain subjects—mathematics, reading, and science—that determined the fates of whole schools. The core challenge was introducing computer science as a new subject into a system of tremendous inertia for which there was no structural policy incentive to adjust or make room for something new.

The year after the STEM Education Act of 2015, the president launched the CS for All initiative. It was a collaborative effort across multiple federal agencies designed to make computer science one of the foundational literacies of American public education. The nod to democratizing computing was built into the title—*for All.* The initiative was a signal to states to begin developing policy structures to incentivize the inclusion of computer science into K–12 education. The initiative catalyzed two efforts at the same time. One was the development of state-level education policies to incentivize the deployment of CS education. School districts needed a reason to offer CS courses. If schools were not being held to account on whether students had CS courses, there was no incentive to put added pressure on already stressed American public schools. Without either an incentive structure or a punitive mechanism, CS education was simply not a priority.

The other signal offered by the federal call was to develop CS courses themselves. Computer science is a relatively new subject. Literacy, for example, has been the bedrock of public education in the United States since the founding of the first public school, Boston Latin, in the seventeenth century. The National Council of Teachers of Mathematics, the professional association of math teachers, was founded in 1920. It is more than one hundred years old. The first electronic programmable computer, the ENIAC, was invented at the end of World War II in 1945. The first computer science department was formed at Purdue University in 1962. CS simply has not been around that long as an academic discipline. The Computer Science Teachers Association, the equivalent professional association for CS teachers, was founded in 2004. Until the Obama administration in 2016, there had been no universal attempt to make the subject broadly accessible to all K–12 students. As such, the second effort set in motion by the CS for All initiative was to determine what ought to be taught in computer science courses to students in their various grades.

The CS for All initiative signaled that computer science had become a national priority, but states have autonomy over American education. It was up to the states to enact the policy incentives to push local school districts to develop and embed computer science courses in public K–12 education. In Georgia, for example, the state legislature passed Senate Bill 108[14] in 2019, which mandated that by the 2023–24 school year all public high schools in the state offer a computer science course. While getting the president to say in the State of the Union that CS was a national priority was hard-fought, declaring that computer science is important was a relatively easy thing to do. Similarly, the acknowledgment of the importance of CS education for the economic development plans of a state is relatively easy to see. For example, when the governor of Georgia is building a strategy to position the state as one of the tech hubs of the United States and lure tech companies like Google, Microsoft, Salesforce, and Rivian, he needs to also be able to say that Georgia has a tech-savvy workforce. That aligns with the push for universal CS education across the state. That is the simple connective tissue between the federal initiative and the state legislative mandate to push the Georgia Department of Education and legislature to act.

The challenges at the school level of introducing a new technical subject like CS, however, are profound. Tackling those challenges is the substance of the ongoing effort to democratize computing. It is analogous to President Kennedy's declaration in 1961, when he challenged the nation to put a man on the moon in a decade, "not because it is easy, but because it is hard." There are slightly more than five hundred public high schools in Georgia that serve approximately five hundred thirty thousand students. In 2019, when SB 108 was passed, there were thirty-three certified CS teachers in the entire state. While the legislative mandate was clear, the state simply did not have the infrastructure to enforce it.

The nation's challenges with mathematics education are legendary and do not warrant further navel-gazing here. One aspect of the challenge on the teacher side, however, is that people with advanced mathematics skills likely have several occupational options that are far more lucrative than being a public school teacher. Crudely, but candidly, there is hardly a fiscal incentive for anyone with advanced technical skills to choose to be a public high school teacher. The choice is more one of cherubic commitment to young people and personal satisfaction than lifestyle advancement or financial security. That challenge is only magnified in computer science, especially in the first two decades of the 2000s. The surging demand for software engineers and computationalists of nearly any kind offered a path to the working American rich. Why would someone with those skills choose to go into public schools fraught with ongoing battles for living wages in major cities and perennial struggles over culture wars that play out in the classrooms? Today, in 2025, the average starting salary for a CS teacher in Atlanta, the tech hub and capital of Georgia, is between $50K and $68K per year. For a software engineer in the same city, a total compensation package averages between $190K and $270K. The median home price in Atlanta is approximately $430K. Teachers are part of America's working poor. These are structural barriers to the fulfillment of the SB 108 legislative mandate. While the challenges facing public school teachers in nearly every major American city are increasingly acute, it is particularly true for CS teachers. Who will the CS teacher corps be? One of the fundamental assumptions of democratizing computing is that there is a teacher corps that can be built to bring the subject to all students. That may not be true. Hence, we may need to redefine what a teacher is and from where the knowledge about CS comes. Not only is there a fundamental challenge in building the teacher corps, but more specifically, the difficulty lies in building it in such a way that the most challenged schools—those that tend to serve Black students—are not seen as undeserving of rare CS teachers because their crises in math and literacy supersede any need for an educational luxury like CS.

One policy approach the democratizing computing movement pursued was to make CS a state-wide graduation requirement for public high schools. The logic behind this policy angle and the end goal of making it universally available was clear. If no student can graduate from high school without CS, then all students who graduate would have some instruction in CS regardless of race, gender, geography, or economic status. The effects of the Obama administration's clarion call ring out here too. Almost immediately, Nevada and South Carolina passed laws in 2017 and 2018, respectively, to become the first states to require computer science for high school graduation. There are now ten states where it is a high school graduation requirement. These kinds of policy changes are the quiet levers that affect hundreds of thousands of students in efforts to bring new disciplines into the American education framework. They are necessary steps along a long policy path that starts at a federal declaration and ends in the individual classrooms of teachers and students.

In American public schools, however, where segregation has been a founding and defining feature, it goes without saying that just because CS is a high school graduation requirement, that does not mean all CS courses are created equal. They are not. We take it for granted that basic literacy is a high school graduation requirement in American public schools too. In 2019, before the pandemic, according to the National Assessment of Education Progress (NAEP), only 37 percent of American twelfth graders were proficient in reading. That same year, less than 25 percent of Black twelfth graders in the United States were proficient in basic reading. In South Carolina, in 2019, although CS was a high school graduation requirement, only 14 percent of Black students were proficient in reading. The challenges to democratizing computing raise a fundamental question: What, to the student who cannot read, is computer science?

These kinds of fundamental education inequities in American public education are built-in barriers to the goal of democratizing computing education. In this era of AI and the assertions that it is the cavalry coming to save the day, the lessons learned and not learned from the decades-long effort to democratize computing become even more important. Not only are the policy efforts to democratize computing fraught with structural challenges, but the specifics of course content are a challenge as well. The mandates for CS as a high school graduation requirement are built on the thesis that every student must do it—that they do not have a choice. They must make special arrangements to opt out; otherwise, they will not graduate. Other states took a more patient approach that mandated that every school offer a CS course, but not that every student must take it to graduate. That is built on the opt-in thesis. The course is there, and you can take it if you choose. Two fundamentally different approaches that have very significant effects on which students end up taking computing courses. Just like math, if you give American students the choice to opt in, the majority will not. The United States is likely one of the only developed countries in the world where people confidently and publicly declare, "I'm not a math person," almost as a badge of honor. Even without the tomes of evidence and data that exist, if we close our eyes and think of who the students are that would predominantly opt out of CS courses in high school, Black and Brown students and girls of all races and ethnicities top the list. We see it culturally as the purview of white and Asian men. If students have to jump through hoops to opt out, however, you have a much broader rate of participation. The opt-out thesis is also predicated on the basic assumption that everyone can do it, that it is not a rarefied specialty for the supposedly gifted and high-achieving few. While these are philosophical policy differences, the second part of the CS for All initiative was to develop the courses that ought to be taught.

Whether students opt in or out, what is a high school–level computer science course, and what are the most effective ways of teaching it? Computer science is a uniquely fast-paced field that renders skills, programs, and techniques obsolete in remarkably short periods of time. By contrast, teaching the fundamental theorem of calculus that relates integration and differentiation has remained largely intact since the late seventeenth century. The relevance of President Obama's 2015 STEM Education Act reveals itself again. The question of what, and how, to teach computer science is an entire educational field. I was a program officer at the National Science Foundation during this time. Explicit in the STEM Education Act was that computer science would be officially included in STEM education efforts at agencies like the NSF. What that enabled the NSF to do in conjunction with the Department of Education was to fund national-scale research to determine the most effective content and pedagogies for high school–level computer science. At that point, and it is still the case that the most standardized version of high school–level computer science was offered by the College Board in the Advanced Placement Computer Science course. The fact that it was an AP course is an indication of who the class is designed for and the assumptions that it is built on. It is an opt-in class and, as AP classes go, students need recommendations to enroll. Here, too, given the patterns in American public education, that means Black students would be the least likely to be on the tracks that lead to AP Computer Science and the least likely to be recommended by their counselors and teachers for advanced anything. And the lack of representation creates a negative feedback loop where Black students don't see other Black students in those classes and then make the individual choice to opt out. Often, the criteria for any of the AP classes, especially technical classes, is at least minimum proficiency in eighth- or ninth-grade algebra. Thirteen percent of Black students in the United States in 2019 were proficient in basic algebra—thirteen out of every hundred students.[15] That is a structural barrier. As such, in 2020, only 3.5 percent of all the students who took the last AP CS A exam—the original Advanced Placement Computer Science course—before the pandemic were Black. Girls made up 25 percent.

These are profound structural challenges that set the limits of success for the democratizing education movement. Without tackling these larger variables, there is only so much that any singular effort to democratize, make equitable, or equalize outcomes in computing education can accomplish. While the parameters around computer science education are particular, the problem is not special to CS. Compounding the structural issues is that the content of the course itself—because of the nature of computer science—remains in flux. The AP CS A course was first offered in 1984. In 1984, IBM was the fifth most valuable company in the world and used Pascal as the primary language to program IBM PCs and compatible systems. It was also the language taught in many undergraduate introductory computer science courses at universities across the country. Pascal was the language students learned in AP CS A from 1984 to 1998. As the field evolved, Pascal was rendered obsolete. From 1999 to 2003, AP CS A was a course in C++. It reflected a shift in the CS community's opinion of the most useful introductory language for students. These curricular decisions were the result of an ongoing dialogue between the Association for Computing Machinery (ACM), which is the professional association for CS faculty, the College Board, and corporate representatives. In 2004, the course shifted again to Java, which is the language that is still taught now. The increasing capabilities of AI-enabled tools, however, are raising questions about the utility of fundamental programming languages altogether. Gemini, ChatGPT, Claude, and other AI tools can produce code in several languages based increasingly on just text or speech instructions.

Because AP CS A is explicitly a course designed to learn programming, in the education world of social and cultural constructs, it is viewed as a class for kids who want to be computer scientists. In addition to the actual barriers, many Black students and girls were simply put off by the boyish persona of the class. Coincident with the launch of the CS for All initiative, work was underway to find a more accessible and welcoming AP computer science course. Within the computing education community, there was ongoing discussion about the utility of the AP CS

A course. The premise behind all the College Board's Advanced Placement courses is that they are supposedly college-level courses. They are scored on a range of 1 to 5, 5 being the best. Students who receive scores of 4 or 5 are often exempted from those introductory courses in college. Of course, at this stage of American life, all parents with children involved in competitive college education know that those courses may or may not be college level, but they are definitely GPA boosters. They are weighted high school courses, so students who take them often have GPAs above 4.0, which ranks them above otherwise perfect students who don't take them—in many cases, Black students who weren't on the right track or simply were not recommended for the courses. The effort to develop a new and more welcoming AP CS course was responsive to this reality. It too was operating under the *for All* clause of the federal CS initiative and was designed to broaden access to computer science beyond predominantly Asian and white boys.

The Advanced Placement Computer Science Principles (AP CSP) course was introduced for the first time in 2016. It was timed to launch with President Obama's CS for All initiative. A critical feature of the class is that it was designed so that students did not need to have previous programming experience. Previous programming experience was an unspoken and informal requirement of the existing AP CS A course and was a structural barrier. While personal computers, tablets, and smartphones are virtually ubiquitous today, just a decade ago that was not the case. Many students, particularly Black and Latino students, did not have computers at home, nor did they have parents who worked with computers—a sad fact that became abundantly clear during the pandemic. They did not have what Jane Margolis, one of the nation's leading scholars on equitable computing education, calls "preparatory privilege."[16] Only a certain subset of American high schoolers had access to any form of computing by their junior year in high school. For all the obvious reasons, that subset correlated with highly privileged, mostly white and Asian students whose parents brought those experiences to them in some form or another. Most Black students, and poor students of all races and ethnicities, simply did not have any experience programming, which rendered them essentially ineligible for the only available Advanced Placement Computer Science course at the time. The door was quietly closed to them for one of the main entrances to postsecondary computing education.

When the AP CSP launched in the 2016–17 academic year, it had the largest number of students enroll of any inaugural launch of a new AP course in College Board history. More than forty-five thousand students took the course. It was clearly tapping into surging interest among students. Many more students than just those who had experience in programming were interested in it. By 2016, "Google" was already a verb in the American lexicon. Facebook was twelve years old, Twitter was ten, and Snapchat was five. *Minecraft*, the second-most popular video game of all time with more than 300 million copies sold, was launched in 2011, and *Grand Theft Auto* launched in 2013. By 2016, American young people of all walks, races, and ethnicities were bombarded with cultural signals of the importance and application of computing. The AP CSP course was, and is, designed to tap into that. It is designed, as the title implies, to introduce students to the principles of computer science. In addition, it examines the interface of technology and society. It is designed to provide some of the exposure, experience, and skills that other students came to organically through their preparatory privilege. The jury is still out about the impact of the course. Questions about its rigor raised by computational purists persist. They argue that since students are not learning a specific computer language, the course is subpar and not rigorous enough—a common critique of efforts to expand access. Whether it translates into more students enrolling in computing-related disciplines in college is another outstanding question. It is virtually impossible to attach causality to a single course—because a student took *x* class, they choose to study *y* major. Regardless, there is one irrefutable fact: The offering of this course has introduced tens of thousands more students to introductory computing than would have been the case without it. In just four years, by 2020, the number of students enrolled in the course had nearly tripled.[17]

There are many ways to determine the efficacy of the social movement to democratize computing education. There are obvious questions: What does winning look like? How would we know if we did? Depending on the interests of those asking, the answers will be different. In the computing education and policy community, the metrics can get granular. How many Black students, students of color more broadly, and girls are taking the AP CSP course? What is the increase from 2016 to now? The abiding race mantra, "you need to see it to be it," inspires questions about how many more students feel that they can see themselves as computer scientists.

It is a question about affect and self-efficacy. Have we broken the imaginative link that binds the image of white and Asian boys to computing such that there is room for others in our own imagination? As a result of this effort, has there been an increase in the number of Black students who are enrolled in undergraduate computing majors at the college level? Given that much of this entire effort began with the surge of Big Tech and the implications of the homogeneity of the most advanced technical workforce, have there been any changes? Has the number of Black tech workers at the Big Tech companies changed in any material way? One could argue, given the unmitigated power that those companies seem to now hold, that this may be one of the most important metrics of whether the democratizing computing effort has been successful.

The most salient point, however, is that the work is unfinished. In the face of shifting attention to AI and a federal retreat from the foundational philosophy of broadening participation in computing, which is the guiding thesis of democratizing computing, we face real risks. The ascension of AI and its increasing dominance in the STEM education world and in education more broadly is exciting and surely offers opportunities for accelerating learning and the closure of educational gaps. It may very well be that AI-enabled tools can make tailored individual learning more effective. It may be that tailored dynamic instruction with AI-enabled tools can make the work of teachers more efficient and accelerate the efficacy of student learning. It may also be the case that AI may be the technical tool that can finally help close the persistent racial gaps in student achievement that are a hallmark of American education. To all of that, perhaps. It all remains to be seen.

Democratizing computing was born in the Obama years, when the President of the United States was pushing for a collaborative effort between the nation's premier basic research agency and the Department of Education to make computer science a new literacy for the nation and to make sure that all students had access to it. Now, the federal establishment is closing the Department of Education and starving the National Science Foundation of its ability to build a research agenda to not only continually improve CS instruction but also better understand the most effective means of bringing students in from the margin. AI in education is exciting, but it is merely next. It will have to confront the ghosts that continue to haunt the democratizing computing movement.

PART II

INVERSE CORRELATION: POWER AND COURAGE

CHAPTER 3

THE EXCEPTION AND THE PATTERN

Exceptions require patterns to exist. The logic is simple and straightforward. For someone to be exceptional, there must be some norm, or some pattern, against which that person has distinguished themselves. You cannot have the former without the latter. At the frontier of the STEM education world and the tip of the tech sector that is pushing AI technologies forward, the relationship between the exception and the pattern for Black people is extreme. The ability to play at the frontier of this new technology era requires a kind of technical training, education, and experience that is out of reach for most Americans. In the Black world, however, the distance between the patterns of educational access and performance in technical subjects, on the one hand, and the requirement for participation, on the other, is a chasm.

That distance between the preparation required to participate at the frontier and the patterns that define Black students' STEM education is the fault line that may ultimately lay the foundation for digital slavery here at the end of the first quarter of the twenty-first century. The significance of this gap between the pattern of Black STEM education and the requirements to engage at the front edge of the tech sector has also been obscured by the marketing of the tech sector itself. Their perpetual highlighting of exceptional Black people and use of them as evidence of progress is a distraction from the fact that very little progress is happening in breaking the patterns that make them exceptional in the first place. That obfuscation of the truth—focusing on the exception to lure attention away from the pattern—is a trap. It offers some of the most powerful companies in the history of the world a place to hide with respect to the issue of race and the tech frontier.

For a decade, from 2014 to 2024, the era of tech-corporate diversity was high-end modeling season for exceptional Black people. The diversity annual reports, brochures, and lobby walls of tech offices were filled with their gleaming pictures. Huge, high-quality photos and videos of successful, smiling, beautiful Black people who loved their jobs, were committed to some company value, and represented deep-seated corporate commitments to diversity. Slogans such as "It is not just good for business; it is the right thing to do" were everywhere, like a spiritual incantation. The quiet marketing narrative was that these exceptional people had successfully made the pilgrimage from social obscurity on the margins—the abyss overflowing with underrepresented, underachieving, at-risk minorities—to the tech promised land. They were images of hope and feel-goodness, warm cookies and cappuccinos, and the tech-corporate nod to a more inclusive work environment and country. They were highlights of exceptions. The patterns from which they were exceptions, however, are always lost, absent, or erased in the telling of the exceptional Black narrative.

This season in the tech sector was also notable for its altruism—particularly toward young people. There were enormous investments of talent, time, and resources dedicated to efforts to expand the tech workforce and make it look more like the broader American population. There were legions of programs designed to expose, engage, and enrich students' access to computing. Here, too, the brochures of these programs were laden with images of young Black students, their faces alit by computer screen light, with whatever programming they were engaged in having completely arrested their attention. Typically, over their shoulder would be standing a branded tech employee fulfilled and glad to be of service, having answered their own question—"How can I help?" The subtext of the marketing of these programs is the sentiment that there, but for our sponsored tech exposure program, goeth a young Black potential software engineer. It is an emotional plea. It is a plea that resonates with the American psyche: the rich giving to the poor; white saving Black. It resonates with the Christian Samaritan: If I do not stop to help this child, what will happen to him? It resonates with popular marketing theology: For just seventy cents a day, you can help young Amadou get some shoes. It is a plea to charity.

It is as difficult to judge the sincerity of broad efforts as it is to judge the sincerity of an individual. The former is merely an aggregation of the latter. What is true is that these efforts celebrated exceptional student beneficiaries of tech altruism. They champion individual students who, through the tech program, some luck, exposure, and their own talent, were able to make it. They made it to the program, if not to further success. But more importantly, they were exceptions to the patterns from which they were plucked. Highlighting the emotional narrative of these successful individual students could be viewed as a distraction to keep our attention averted from the actual problem that entrapped them in the first place. Solving the challenges of an individual is a far easier task than tackling the structural forces that create the challenges individuals face. Over-indexing on the individual story and leaning on the heartwarming narrative provides a hiding place to say that work is being done, commitment exists, students are being helped, progress is being made, all the while the actual problem remains unsolved. The standard mantra in reporting about this body of work is, "We're making progress, but there is more work to do." In some ways, the focus on the exception is an absolution for neglecting the larger forces that create the patterns. It could be that the exaggerated celebration of the exceptional cases as proof of progress is merely a veil to hide the real lack thereof. It may not be that malevolent. Perhaps.

I may or may not be an exceptional person, but I am an exception to the pattern of Black people in STEM education. It was glaringly apparent in that high school in the Bronx. There, the exception and the pattern came face-to-face. The principal's racial transference argument highlighted the confrontation and relationship between the exception and the pattern. The subtext of his message was that my ability to achieve as a Black person ought to inspire and indicate possibility for the students' achievement. It may have, but what is it exactly that we're asking of Black students—fourteen- to eighteen-year-old children—trapped in these patterns as they were? Virtually none of the thousands of students currently in that Bronx school, or the thousands in the seasons before or since, had passed algebra, and the school didn't offer calculus. They existed in, were constrained by, and ironically defined the pattern that trapped them. To break that pattern, what we're asking of them is to essentially be superhuman. We are asking them to be something they'd never seen, by way of courses they weren't offered, while resisting the advice of guidance counselors guiding them to a sharecropping life. More powerfully, we're asking them to construct an identity for themselves that can withstand the withering and relentless assault of doubt about their merit, that is the oxygen of the world we're asking them to enter. The famed psychologist Angela Duckworth and her book *Grit: The Power of Passion and Perseverance* are often referenced to suggest that these students need to dig deep.[1] They need to lean on their passion and persevere despite the obstacles. Yes, of course, we all need grit. We need to be able to embrace delayed gratification and not bend to the first wind or break in the storm. That narrative resonates with human heroism and is the stuff of real success. Overcoming extraordinary odds and winning is part of the American narrative; it is the Hollywood version of America. However, it is not the formula for breaking patterns. The transformation of institutional structures is the thing that breaks patterns, not remarkable individual heroism.

A dimension of this ask of these particular Black students harkens to a tradition that we do not introduce in the staid discussions of access to the tech enterprise and STEM education. It is a confluence of two world experiences that seldom meet. In many Black churches, when pastors discuss faith, they anchor it to the origin story of Black people in America. They cite its definition in Hebrews 11:1—"Now faith is the substance of things hoped for, the evidence of things not seen." They talk about it in terms of Black people's yearning to be free from bondage. The pastors remind us that we were enslaved for 246 years. That is approximately ten generations of people. In ten generations of brutality and barbarism, the reality of a pre-slavery existence can morph from memory to myth. After nearly two and a half centuries, Black people yearning to be free in America were dreaming of something they'd never seen. They were drawing on something of which living memory had long since died. They were hoping for something there was no evidence to suggest was real or could be, yet their belief in freedom was steadfast. Despite every effort to smother it, their faith remained untrammeled. There is an element of that tradition embedded in the ask to these students. To be clear, participation in the STEM enterprise is not analogous to freedom. That is not the comparison. What we're asking of them is to envision a world for themselves that they've not seen and have been intentionally and generationally denied. That is an extraordinary ask of the imagination and fortitude of a fourteen-year-old. That isn't grit. It bears repeating that the first whole generation of free Black people in American history was born after 1964. They were teenagers in the late 1990s. Their parents were born in the 1930s and '40s, around that pivotal moment in the country, in its transition from an apartheid to a free state. That first generation was born into an America where freedom was a brand-new garment.

The traps in the Bronx, in Brooklyn, in Harlem, and in every other enclave that concentrates Black people in major urban centers across the country have their own history. The current attempts in public education to erase the history that plainly identifies the responsible fingers of racism don't extract that history from the DNA of these students or the communities from which they come. They know. In James Baldwin's "A Letter to My Nephew," from 1962, he lays out the nature of the ask:

> This innocent country set you down in a ghetto in which, in fact, it intended that you should perish. Let me spell out precisely what I mean by that for the heart of the matter is here and the crux of my dispute with my country. You were born where you were born and faced the future that you faced because you were black and for no other reason. The limits to your ambition were thus expected to be settled. You were born into a society which spelled out with brutal clarity and in as many ways as possible that you were a worthless human being. You were not expected to aspire to excellence. You were expected to make peace with mediocrity.[2]

That contractual peace with mediocrity is the pattern. It is revealed in every data point that outlines Black educational achievement in America. The platitudes about gains, so common in school district leaders' progress reports, do a poor job of hiding the absence of mastery. Mastery is the portal to excellence and the antidote to mediocrity. If these students are not expected to aspire to excellence, then pushing them and building a structure to achieve mastery is unnecessary. The logic is simple and straightforward. Ironically, mathematical logic that students typically learn in geometry, which comes after algebra, helps us see the relationship plainly. If p equals excellence, and q equals mastery, then the conditional statement reads, if p then q ($p \rightarrow q$). If there is excellence, then there is mastery. It offers the condition that for there to be excellence in educational achievement, there must be mastery of foundational subjects. Said another way, excellence is conditional. It only exists if there is mastery. The conditional statement can only be true if, and only if, the contrapositive statement is also true. If not q, then not p ($\sim q \rightarrow \sim p$). If there is no mastery, then there is no excellence—a relationship similar to struggle and progress. Without mastery of foundational subjects, it is impossible for these students to achieve educational excellence. The logic remains clear. Parents, teachers, and students themselves know that whatever the platitudes and ribbons may reward, if there is no mastery, there cannot be excellence. Excellence requires mastery. The fact that nearly none of the students in that Bronx school had passed algebra suggests that there was certainly no mastery, and therefore, excellence in any STEM domain was impossible.

Under that basic logic map, the racial transference argument does not work. It is not enough to be merely inspired. There will be students whose indomitable spirit renders them unconquerable. There always have been, and there always will be. In his letter to his nephew, James Baldwin references that spirit too. "Know whence you came. If you know whence you came, there is really no limit to where you can go." That is the tradition that pastors in Black churches call on and the spirits that they conjure. Black people were American slaves—a people strangled by the American origin story, yet, as Langston Hughes declared, still "sing America."[3] Behind Baldwin's counsel to his nephew was that yearning of a beleaguered people. He is alluding to Psalm 121. You can hear it in his language. "I will lift up mine eyes unto the hills, from whence cometh my help. My help cometh from the Lord."

These students were trapped in the Bronx, in an institution that, if not designed for them to perish, was doing nothing to prevent them perishing. Where ought they look for help to break the pattern? So many avenues of escape were cut off to them that they could easily suffer atrophy of ambition. It would not be an unreasonable outcome. That is what it means to be coaxed into making peace with mediocrity. You settle for the path you can carve with the tools you've been granted. For the individual Black student, breaking free of that pull to mediocrity is more than grit; it is defying a historic American expectation of amounting to not much. These considerations of what it really takes, of what is really being asked of the spirit of a child, reveal the challenge hiding behind the veil of the frightful statistics. They ground the whole endeavor more deeply in the American story.

These parts of the patterns that relegate Black students to the margins of advanced STEM education are seldom considered when edtech companies and large tech corporations suggest technical solutions to social problems. They are lost when exceptional students are highlighted as evidence of progress made. Exceptional students who make it out from these circumstances are used as false proof positive of the American story at work—of individualism, of pulling yourself up by your bootstraps. If you work hard enough, anything is possible. If they chose to succeed, they would. Exceptions don't make those claims true. They merely make them possible. The truth lies in the patterns. For Black students in many schools, the will to succeed and work hard is hardly a guarantee of success. Even for those students who aren't lulled into accepting mediocrity, breaking free of the pattern is not assured. If the courses aren't offered, then even their best isn't good enough. Taking all that is available and doing well still leaves them falling short of the minimum qualifications of entry to the high-level STEM enterprise. If calculus is not offered in their school, for example—as was the case at that school in the Bronx—they are structurally ineligible to be competitive at elite technical universities. Despite all the populist egalitarian arguments to the contrary, those elite technical institutions remain the primary conduit to the frontier of the tech enterprise. If students don't have calculus in high school, they are virtually locked out of entry to elite technical institutions. They are structurally ineligible through no fault of their own.

In New York City, where I grew up, there are the famed tested-entrance public high schools—Brooklyn Technical High School, Bronx High School of Science, and Stuyvesant High School. These are schools where students must take an entrance exam in order to attend. Stuyvesant is widely regarded as the most elite public high school in the city and among the best and most rigorous high schools in the country. It exists within the same New York City public school system as that school in the Bronx where I was sent to inspire in the late 1990s. The NYC public school system is massive. It is by far the largest single public school system in the country. It serves nearly a million students. It is more than double the size of the second-largest school system, the Los Angeles Unified School District. While enrollment in NYC public schools has been declining in recent years, there are just shy of eighty thousand students in the eighth grade alone. Given a school system of that size, patterns are revealing.

They highlight with irrefutable evidence the relationship between exceptions and patterns.

For students to enter the tested public high schools in New York, they must sit for the Specialized High School Admissions Test (SHSAT). It is arguably the most competitive exam students in New York City public schools take. The high stakes and pressure are extraordinary. Over the last several decades, whether spoken out loud or not, everyone knows that these schools are the portals to elite postsecondary education at the nation's most prestigious colleges and universities and, ultimately, to the top of American life. Notwithstanding the increasing dismantling of that broad social contract in this era of American anti-intellectualism, these schools stand tall at the gate. They offer among the best chances for academic and social success, and that remains absolutely true for entry into the high end of the tech frontier as well. All three of NYC's top schools—Brooklyn Tech, Bronx High School of Science, and Stuyvesant—are founding members of the National Consortium of Secondary STEM Schools (NCSSS). They constitute the most elite public high schools in the United States that focus on STEM education. The pressure to get into these schools and on that path starts early. Kaplan offers individualized tutoring for the SHSAT for $3,600 for thirty hours. There are summer prep camps that cost more than $4,500. Some families pay tens of thousands for private tutoring starting in the sixth grade. The price of the ticket is always an indication of the value of entry.

Arithmetic, algebra, and geometry are the core math subjects tested on the SHSAT. Arithmetic is foundational numeracy, understanding numbers and their sequences, counting, and basic numerical operations. It is something that all students begin learning when they are learning the alphabet and how to spell. It is that fundamental. While the significance of geometry is arguable, algebra is ever-present. The origin of the word is Arabic, *al-jabr*. It means the reunion or completion of broken parts. It is the branch of mathematics that deals with abstract systems and the equations that shape them. It is the basis of the ability to solve linear and multilinear equations. But most importantly, it is the introduction of the *variable* as an abstract representation of a value. That is the cornerstone of any STEM endeavor thereafter.

For people who do not think about such things, a basic example of an algebraic expression is $x + 5 = 10$. To a student seeing this for the first time, the obvious question is "What is x?" They would surely know by the ninth grade that $5 + 5 = 10$; therefore, $x = 5$. But the concept is that x is a variable to be solved for. In more complicated relationships, they wouldn't know a priori that it is equal to 5. Higher-order polynomials don't have obvious solutions. Third-order polynomials, for example, $f(x) = ax^3 + bx^2 + cx + d$, don't have obvious solutions. You must solve them. That is the progression of algebra. In the $x + 5 = 10$ example, to solve for x, you cannot "break the equation." You must treat both sides equally. Subtracting 5 from both sides isolates the variable x on one side of the equation and allows you to solve for its value: $x + 5 - 5 = 10 - 5$. Therefore, $x = 5$. This isn't a math lesson; rather, it is an attempt to illustrate what it is that so many tens of thousands of Black students are struggling with behind the broad claims about their proficiency in algebra or lack thereof. It is important to remind ourselves what those foundational concepts are and how important they are to the STEM enterprise. Of course, the complexity of the relationship between variables and what they represent can become infinitely more complicated. The inability, however, to grasp this foundational relationship between a variable and its value is a debilitating handicap. Mastery of this basic concept of a variable and its behavior in relation to other variables is the cornerstone of the systems of equations that unlock the mysteries of the scientific enterprise. There is no escaping that beautiful truth.

New York City Public Schools is approximately 25 percent Black students. That proportion has been decreasing in recent years as the number of Asian and Latino students is increasing rapidly. However, each year, of the approximately eighty thousand eighth graders, roughly twenty thousand of them are Black. In the years 2021, 2022, 2023, and 2024, the number of Black students admitted to Stuyvesant was eight, eleven, seven, and ten, respectively. The pattern was the same before the pandemic. Between 2014 and 2020, the number of Black students admitted each year ranged from a high of thirteen in 2017 to a low of seven in 2014 and 2019. Of the eighty thousand or so Black students who traveled through eighth grade in NYC public schools during those four years closing out the first quarter of the twenty-first century, thirty-six of them were admitted to the most elite high school in the largest public school system in America. The freshman class at Stuyvesant is approximately 850 to 900 students. So, to be clear, if Black students were approximating their percentage in the system, there would be nearly 225 Black students admitted each year—approaching one thousand over a four-year period. Therefore, those thirty-six students are, by definition, exceptions to the pattern.

While having mastery of algebra does not guarantee admission to Stuyvesant, not having proficiency guarantees rejection. Given the centrality of algebra on the SHSAT, students who aren't proficient in algebra are structurally ineligible. No matter what they do or how inspired they may be, in schools where it is poorly taught, or not taught, or where there is no expectation of their passing, they cannot gain access to the most elite school in the city. Symbolically, the gate to the frontier of STEM education is not just closed; it is locked. That is the pattern. In the decade between 2014 and 2024, which coincided with the surge of Big Tech and its diversity decade, 102 of the roughly 220,000 Black eighth graders in New York City were admitted to Stuyvesant High School. Of course, the argument is not that Stuyvesant and schools like it in the NCSSS are the only path to the frontier. Clearly, that isn't true. However, in the context of access to the research and development of artificial intelligence at the cutting edge of this new technical world, and access to the emerging unbridled power associated with that, elite technical education matters.

It is also worth noting that this pattern has historical significance. The year 1976 was the bicentennial anniversary of the founding of the United States. The nation marked its two hundredth birthday. That year, four years after I was born, and just twelve years after the passage of the Civil Rights Act ending legalized racial discrimination in America, 10 percent of the students admitted to Stuyvesant were Black.[4] Over the next fifty years, that percentage has steadily decreased to what it is today, hovering around 1 percent. We celebrate the 250th anniversary of the United States in 2026, and there will likely be only four or five of New York City's approximately twenty thousand Black eighth graders admitted to Stuyvesant. Keeping with that trend, the first year in post–Civil Rights Act America, when there are no Black students at all admitted to Stuyvesant High School, will be a milestone year. It appears to be coming soon. That year will mark a new beginning—a second fight for integration, a second season of pioneering Black fourteen-year-olds trying to cross the Rubicon into a land alone. Ruby Bridges will send her spirit to hold their hands and offer calm assurance that we've been here before.

There is no shortage of American voices—typically white, Asian, and newly arrived immigrants—that proclaim that Black students don't achieve because they don't work hard, they are lazy; they don't have a culture that values education; they are driven by immediate gratification; they don't have the intellectual discipline or disposition for the rigors of advanced technical education; their parents are careless, clueless, and undisciplined; they blame the system for their own failure; they see themselves as victims; they don't take responsibility for their behavior; they only progress because of the cherubic charity of white people offering them affirmative action that short-circuits merit; they get what they deserve; et cetera.

As we celebrate the 250th anniversary of the country, those voices are only emboldened. Irony would have it that the President of the United States, during this national milestone, says unabashedly that Black people eat dogs, and that African countries are shitholes.[5] He called for the killing of innocent Black teenagers in New York when he was younger.[6] On the day after Martin Luther King Jr. Day in 2025, he issued an executive order[7] explicitly revoking the 1965 Executive Order 11246[8] issued under the Johnson administration that prohibited employment discrimination by federal contractors and subcontractors, laying the groundwork for the Equal Employment Opportunity Act of 1972. He welcomed back the villainous spirit of segregation and discrimination. The EEOC was the federal mechanism by which the American public could fight discrimination that is so deeply embedded in American culture. It was issued one year after the signing of the Civil Rights Act of 1964. Now, once again, racial segregation and discrimination and all that they imply—including in education—are no longer barriers to doing official business with the federal government of the United States. The door is open again for the government itself to return to the discriminatory days that warranted President Johnson's executive order in the first place. President Trump's scorn and blatant contempt for Black people is quite literally returning the country to a time of segregation—a time when the rules created the patterns. Those voices, so quick to demonize Black students for their underachievement, blithely overlook historical reality and don't notice that their ankles aren't chaffed by government-issued slave chains.

Those blustery critiques of Black students are often offered with a measure of self-proclaimed and self-righteous objectivity as insolation from their racist central thesis. It is a laundering of sorts. They use the very data I offer here as evidence of their professed logical conclusion that Black people ought to be rendered as digital slaves in a technical age for which they are not well suited. If they cannot do algebra in a world where algebra is the portal to the front of the STEM enterprise, then perhaps they should settle in the back. They should just rise to the level suitable for their mediocre talent and be at peace with that. In the 2016 US Supreme Court case on the consideration of race in college admissions, *Fisher v. University of Texas*, also known as *Fisher II*, Justice Antonin Scalia gave voice to this very sentiment during the oral arguments.[9] He pushed back on the Solicitor General, who was representing the University of Texas, and offered:

> There are those who contend that it does not benefit African Americans to get them into the University of Texas where they do not do well, as opposed to having them go to a less-advanced school, a slower-track school where they do well. One of the briefs pointed out that most of the Black scientists in this country don't come from schools like the University of Texas. They come from lesser schools where they do not feel that they're being pushed ahead in classes that are too fast for them. I'm just not impressed by the fact that the University of Texas may have fewer [Blacks]. Maybe it ought to have fewer. . . . I don't think it stands to reason that it's a good thing for the University of Texas to admit as many Blacks as possible.[10]

Because of the historic milestone in the 250th anniversary of the nation, it is worth citing the origin story of malevolent critiques like Justice Scalia's and the relation they have to the patterns we now see. Given that the President of the United States has offered safe harbor to the unmitigated public slander of Black people from the highest office in the land, it is useful to remember that narratives have ancestors too. The president's sentiment is a direct descendant of the American attitude toward Black people that was the cornerstone of the nation's foundation. In 1858, South Carolina Senator James Hammond offered the full thesis of what our contemporaries allude to but eschew.

> In all social systems there must be a class to do the menial duties, to perform the drudgery of life. That is, a class requiring but a low order of intellect and but little skill. Its requisites are vigor, docility, fidelity. Such a class you must have, or you would not have that other class which leads progress, civilization, and refinement. It constitutes the very mudsill of

> society and of political government; and you might as well attempt to build a house in the air, as to build either the one or the other, except on this mud-sill. Fortunately for the South, she found a race adapted to that purpose on her hand. A race inferior to her own, but eminently qualified in temper, in vigor, in docility, in capacity to stand the climate, to answer all her purposes. We use them for our purpose and call them slaves.[11]

The patterns demonstrated by Black students' lack of access to schools like Stuyvesant and their concentration in schools like those in the Bronx are born of rules and shaped by the cultural expectations reflected in iconic American figures like Senator Hammond, Justice Scalia, President Trump, and others of their ilk. The legal rules of segregation and the social rules of separation that defined public education in de jure form until 1954 and in de facto form ever since are what give rise to these patterns of educational underachievement. The comprehensive conditions, if not plight, of Black communities in New York and in urban centers around the country are the array of variables that undermine successful education. It is worth noting that the Bronx is the poorest borough in New York City, where today nearly a third of all the residents live below the poverty line. That wave of extreme concentrated poverty began in the 1970s.[12] There are tomes of research and armies of academics in the education world trying to solve this problem of the racial partition in American education. The goal of the work is to establish a public education system that offers equally high-quality education for Black and white students in a society that doesn't value them equally and never has. In algebra terms, it is a broken equation. The coefficient of value on Black students is less than and not equal to that on white students. Notwithstanding high-minded academic educational parlance and heavy-handed education policy agendas, that's a tough nut to crack.

This is not a view of fatalism; rather, it is an attempt to fully understand the relationship between the exception and the pattern. It is important to position the goals of a more fair and just STEM education enterprise in the real contemporary and historical context in which they sit. The application of technical solutions to social problems requires that both the technical and social parts of the equation be fully prescribed. Just because AI has come to dominate the STEM education landscape does not make the body of work either apolitical or inoculated from the historical traditions from which its challenges in education arise. The confluence of the conditions, attitudes, and institutional inertia that formed the patterns not only dictate the outcomes for so many Black students, but they also raise the question of whose responsibility it is to help break these patterns. The magnitude of the forces that keep the patterns in place, as well as their historic roots, are often hidden from view when Big Tech companies offer polite, sterilized consideration of technical altruism. CS for All can be a shield to hide behind, allowing for the very narrow approach to a highly prescribed problem that pretends not to see the raging racial battle that motivated its existence in the first place—give everyone a computer, or an hour of code, and hope for the best. Alternatively, CS for All can be a call to arms to offer a new charge and a new problem frame. What pressure could be put on cities clamoring for the distinction of being a tech hub if they don't find a new path to address the broken equation of their public education? Returning to the Christian Samaritan, large companies benefiting from all the allowances they're granted for entry into a city could lead with the question: If you don't stop undereducating these students, what will happen to them? They could adopt the position: If you want me, you must do right by them.

How can these new social agents be more integrated civic partners to help solve some of the fundamental, perennial challenges that limit the impact of their engagement efforts? The extraordinary rise in power of the tech sector has changed the equation. It introduces new normative questions because local-level stewardship of public education is not, on its own, solving the problem and, in the current climate, is sometimes making it worse. The changing power dynamic and the unveiled vitriol from the highest office in the land also introduce new questions. The questions force us to consider whether, in fact, we want these problems to be solved, and who the *we* is. Do we want Black students to rise to the level of technical prowess and take seats proportionate to their numbers around the most powerful tables in the world? Or do we ultimately and quietly aspire to a modern version of Senator Hammond's goal of having Black people be a race inferior, suited for a second-tier life serving the gilded class with Black jobs, and call them digital slaves?

Any attempt to change the structural foundation of US education is a paralyzing prospect even for people who work in the education world. It leaves even the most committed pleading, as Reinhold Niebuhr did, "God, grant me the serenity to accept the things I cannot change, the courage to change the things I can, and the wisdom to know the difference." During the 2014–2024 era of tech diversity and the celebration of Black exceptionalism, tech companies, like so many others, concluded that they would ignore the patterns and accept them as things they cannot change. They accept them as constant variables in a social equation that leaves Black communities permanently entrapped and Black students perpetually locked out of the tech enterprise. Their working assumptions were that it was beyond their purview to try to influence local education in any meaningful way. They, not unreasonably, declared that public education is not the domain of Big Tech, except as a client for their tech tablets and software. They raised the normative question of whether private sector agents ought to be involved in public sector affairs like education—all reasonable. Yes, tackling the patterns is a Herculean task not for the faint of heart. But the involvement of Big Tech in this space, in this era, is different. These are not small, or even large, mission-driven nonprofits trying to change a system. Neither are they socially conscious for-profit companies. These are arguably some of the most powerful institutions in the history of the world at the height of their power—and they made public declarations that equity, justice, fairness, and diversity were foundational to their core values. They are more akin to unarmed nation-states. The argument that public education is beyond the purview of their core mission—to generate technology and profit and maintain accountability to shareholders—is legitimate, of course. But there is a power dynamic here unique to our time. It raises the question: What is the civic responsibility of the most powerful agents of our time? And if they themselves declare that they are committed to creating a more representative and just landscape in the tech enterprise, what ought to happen?

When I arrived at Google in 2018, I was responsible for creating innovative means to partner with higher education institutions and secondary schools across the country to improve computing education. This was part of the private sector side of President Obama's push for CS for All. At the time, Google was responding to its nearly insatiable appetite for computationalists—software engineers and computer scientists. It was also responsive to the national movement to diversify the tech workforce in those years—four years after the Reverend Jesse Jackson forced the tech companies to act. In 2014, Rev. Jackson saw diversity in the tech sector as the next frontier of the Civil Rights Movement.[13] He publicly pressured the leading tech companies to release their EEO-1 reports.[14] That is how the world began to learn just how few Black people were working at the frontier of the emerging Big Tech companies—1 percent in tech roles and 3 percent in non-tech roles at the major companies he surveyed. He brought the weight and reputation the Rainbow PUSH Coalition had at the time to bear. He forced their hands and threatened to make formal EEOC complaints against the major tech companies. Surely with intention, it is the very lever he used to push the tech companies to reveal their data that President Trump's executive order has revoked. Rev. Jackson arguably is singularly responsible for the decade-long diversity era in Big Tech. Up until his pressure, not only did the tech companies not offer diversity annual reports; they hid the diversity data. It was the pressure that Rev. Jackson applied with his coalition, and ultimately, President Obama, that forced them to act. The high-quality annual celebratory we're-making-progress-but-there-is-more-work-to-be-done reports that we've become accustomed to seeing were not born of willing intention, but under intense pressure from a civil rights icon and the first Black president of the United States.

Having been at Google only three months, the Chief Diversity Officer at the time arranged for me to speak on the TGIF stage. TGIF is the uniquely Google platform for dialogue between the senior leaders and the employees of the entire company. Inside Google, it has had a tradition akin to a cultural holy grail. For many in the company, it is a career aspiration to speak at TGIF. For many years after the company was founded it remained one of the few open fora in a corporate space with no script and no filters. In the beginning, Larry Page and Sergey Brin would discuss freely with the whole company their plans and visions and hear feedback in a freewheeling open dialogue. In recent years, as is the case with most good things that require community trust, it has become fraught with leaks and political and social subterfuge. It is significantly scaled down now and not at all an open forum. When I was given the opportunity, however, I was able to take the stage on the heels of Sundar Pichai, the CEO, and discuss my views on the inequities of computing education with no script or prescribed talking points.

I talked about the goals we articulated as a company for achieving a meaningful transformation of the demographics of the computing community of software engineers at Google and in our sphere more broadly. My basic argument was that there were simply not enough Black graduates coming from the elite institutions—from which Google drew the majority of its technical talent—to achieve the goals as described. The rate of increase of Google's demand simply outpaced the production of Black software engineers. All the hyperbole and declarations of commitment wouldn't change that basic fact. It was a problem we could neither recruit nor poach our way out of. If we were keeping our goals attainable, we had to address the supply. That meant we needed to exert our influence in increasing the quality and capacity of institutions to produce more Black students with advanced computing skills. It would only be solvable by changing the production mechanism, which meant getting involved more directly in education. I'll return to the normative questions of that argument, but while on the stage at TGIF, my case was compelling and well-received. It was fact-based and data-driven. The kind of case that is effective in Google world.

After my presentation, the Chief People Officer, Eileen Naughton, pulled me aside. She wanted to chat. We went for a stroll in the Googleplex, just she and I. I'd never met her before. She is wrapped in elegance and power. She speaks softly and clearly, with unmitigated strength. She exudes confidence and grace. We exchanged our introductory pleasantries. I liked her. It seemed mutual. At one point, she said she found my argument compelling. She didn't think she'd ever heard such unscripted truth about these matters told so directly from the stage at TGIF. It was refreshing, she said, but that I pointed out a problem that isn't new; we just don't talk about it so plainly and openly as I just had. It has been plaguing this country for generations. She pointed out, however, that now we have new players and new technologies. She spoke of the Khan Academy and its individualized educational tools and the transformative impact they can have. We spoke about the dependence certain communities have on public schools and the historic and fraught relationship between states' rights and the undereducation of Black children. She even brought up the Kerner Commission Report, sanctioned by President Johnson in the 1960s to identify the structural tenets of racism in America, which very much included education. We chatted about whether the traditional definition of school was adequate for this stage of the early twentieth century and, if not, how we would redefine what the institution is and who or *what* a teacher is. Then she said, just as plainly and calmly as everything else she'd said to that point, "Kamau, if we [Google] really want to make an impact in the education production at this level, as you point out, do you think we should build a university, or buy one?"

I was stunned.

There are a lot of reasons to be highly skeptical of the normative premise of that question. Of course. But what struck me was the power behind it. I'd never encountered that kind of power and assurance. Having worked at Georgia Tech, at the National Academy of Engineering, at the National Science Foundation, at the National Action Council for Minorities in Engineering, at the Stanford Research Institute, I simply had not encountered that kind of power so calmly exuded. University systems and governors of states establish universities. Nations establish universities. In my world of storied higher education institutions, it does not even come up as something that happens at this stage of American history. Georgia Tech was established in 1885, twenty years after the end of slavery. UC Berkeley was founded in 1868, eighteen years after California became a state. People in the STEM education community battling for more equitable outcomes had for years been operating in the policy structure, in the realm of advising and innovation in pedagogies and curricula, in recruitment and in the provision of scholarships, internships, and research opportunities. Community leaders and advocates surely start schools at the local level and many aspiring education champions have started schools, especially in the charter school era. But this was not that. This was one of the senior executives of one of the most powerful and transformative companies in world history calmly contemplating whether to build or buy a university.

It was my first exposure to that kind of power and the expansion of imagination it unlocks. In that world view there are very few fixed variables. Anything can be changed. The heads of some of these Big Tech companies have their own space programs. In Black-circle language, white people steal countries, Black people steal cars. It is a contrast of relative power and the freedom of imagination it engenders. For some reason, it left me deeply contemplative about the nature of the problem we're trying to solve. I wondered about whether the locus of power to solve the problem is collocated with the conviction to do so. There remains a tension in Black communities that screams about how much to enlist others to help solve problems within Black communities. It is one of the ideological differences between Malcolm X and Martin Luther King Jr. That tension has roots in Black people's origin story in America and continues today. In 1964, Malcolm X had recently separated from the Nation of Islam. In his famous address "The Ballot or the Bullet,"[15] he laid out the philosophy of Black Nationalism. It centered on the idea that the segregation and racial hostility that defined American life required Black people to turn to ourselves—to establish our own political and economic philosophy of self-sufficiency as the foundation of our redemption. The premise was that we couldn't rely on white people who offered no historical evidence of being worthy of Black trust. Hence, Black people have to nation-build, which means establishing and relying on our own institutions of business, faith, and education. It raised very fundamental questions about the interdependence of America's various people, but it anchored on Black self-sufficiency and agency.

Whether the strategy was segregation or integration, the conditions Black people found themselves in were not self-inflicted. Malcolm X and Martin Luther King Jr. absolutely agreed on that point. In his interview with Sander Vanocur on NBC in 1967, Martin Luther King Jr. tersely laid out the tension when he was asked essentially why Black people are struggling to succeed while other immigrant groups have progressed so much more.[16] Dr. King offered that Black people were freed from slavery by the Emancipation Proclamation in 1863, but after 244 years of slavery, had nothing and were given nothing. No other immigrant group of any kind from anywhere had been enslaved in America, he noted. None of them had been "thingified," stripped of their humanity and reduced to a thing, the way Black people have been. Yet immigrant white peasants had been given millions of acres of land to settle and resources with which to do so by the US government for the expansion of the American West and Midwest. The destiny manifest in Manifest Destiny was that of the dominion of white people over everyone in their path—the slaughter of Native Americans, the subjugation of Black people, and the trampling of Mexican and Chinese people. That set up their economic base. The foundation of their progress was built on generous handouts from the government. Black people, however, were given the freedom to be hungry, the freedom to be exposed to the wind and rain. Freedom without food to eat or land to cultivate. He suggested that Black people must do all that we can for ourselves for sure, but coming out of slavery, "it was freedom and famine at the same time. And when white Americans tell the Negro to lift himself by his own bootstraps [and overlook their singular responsibility for two and half centuries of slavery]. . . . it is a cruel jest to say to the bootless man that he ought to lift himself by his own bootstraps."

Embedded in Eileen's question was not only the revelation of power but the question of who is responsible for solving the problem of these crippling patterns. Behind what I trust was her absolute sincerity in thinking boldly about a solution to a seemingly intractable problem was the more direct question: Can Google help Black people solve this problem of the patterns? More specifically, in an era when 102 out of 220,000 Black students are admitted to Stuyvesant High School per decade, can Google help carve a bigger path to the front of the STEM enterprise? It is a more narrowly bounded question, but given the current reality, it has the potential to make a significant impact. In a world that is often hidden from Googlers and high-end policy advocates, there are Black people who scoff at the mere suggestion. Expecting Google, a for-profit technology company led by uber-rich white and Asian people, to help solve the problems of Black people is foolhardy on its face. Why would they? is the obvious question. It is not an organization with a social mission, so no group of people ought to have such expectations. In 2014, the year that Google started recording its demographic data, 1.5 percent of its tech workforce was Black. In 2024, that percent had risen to 4.2 percent—and these percentages are higher than they were before 2020, when Google began counting people of mixed race as Black too.[17] Four out of every hundred tech Googlers were Black in 2024. Over that same period, Google's valuation rose from $360 billion to $2.4 trillion. Google doesn't need Black people. Its products are global, and its ad revenue is seemingly boundless. Black people are not Google's problem.

Within that same Black community of thought is the foundational principle: We need to *build our own*. Build our own schools, our own corporations, build the strength of our own communities by turning our attention inward and investing our time and talents into ourselves. That is the path to breaking the patterns. It is the cornerstone of Malcolm X's basic thesis of Black Nationalism. The only way Black people can thrive in America is if we are singularly responsible for our own success. Malcolm X argued there is no evidence in American history to suggest that Black people can rely on white institutions for salvation. "When I lift up mine eyes unto the hills from whence cometh my help, my help cometh from the Lord" is how the refrain goes—or, given that Minister Malcolm was Muslim, Surah At-Talāq, "and He will provide for him from sources he never could imagine." We can only break these patterns by internal self-reliance and divine assistance; that is the bedrock of his argument.

The challenge with that argument, of course, is the very point that Martin Luther King Jr. made. Despite the unofficial American motto of pulling yourself up by the bootstraps, especially popular among white people and successful recent immigrants who think that their spoils are solely the result of their personal merit and hard work, no other group of people in America has had to overcome the very specific and distinct assault associated with being Black in America. Even given that, the truth of the matter is the United States is becoming a much more racially and ethnically diverse nation whose people are increasingly interconnected. The ideal of a multiracial, multiethnic democratic society remains, in my view, the goal of achieving a more perfect union. With that comes increasing interconnectedness and interdependence. Of course, community self-reliance and personal responsibility are the bedrock of social success, but too much internal isolationism is merely segregation for different reasons. It is a concession to a lesser American ideal.

These much bigger issues lurking in the background help give Eileen's question shape. One is normative. Ought powerful Big Tech agents get involved in the problem of trying to break the patterns that keep Black students out of their own ranks and the epicenter of power in modern American life? To date, the answer has been emphatically no. The other is that if they choose to do so, what should they do? These are nontrivial questions to be addressed for sure. The question of whether a Big Tech firm ought to establish a university to increase the number of Black students educated at the highest levels of computing may seem unreasonable on its face to many. It raises obvious questions about the relationship between corporate for-profit interests and academic freedom. The central philosophical challenge pits the idea of education for the sake of education against the idea of postsecondary schooling as merely advanced occupational training—the former yielding educated citizens, the latter well-trained workers. All are big issues that require real engagement. But for the purposes of solving the narrowly prescribed problem of the number of Black students who have access to the leading edge of the current technology wave, the proposition warrants some consideration.

If the pattern we're trying to break is not the entire enterprise of Black student representation in STEM fields but rather simply making the pool of Black students with access to the front end of the tech space bigger, there are some details to consider. Let us assume that the top ten undergraduate computer science departments in the United States are the primary feeder programs to the frontier of the tech sector, where artificial intelligence research and education are accelerating apace. Over the last quarter century, since the beginning of Big Tech, these schools and companies have been intimately connected. These are the big Research I institutions leading the most advanced research and educational developments in artificial intelligence and computing. They are the most closely connected to the tech sector both in funding and in the cross-pollination of faculty. Student internships, research summers, and first jobs also flow in a circular current between these companies and the top schools. Many of the research centers and faculty at these institutions are funded by tech sector philanthropy—both the formal foundations of the corporations and by the private family foundations of many of their senior executives who are now among some of the richest people in the world. The founders and leaders of many of the tech companies are alumni of these elite technical institutions, and many faculty share joint appointments. Their names are engraved on buildings and centers and emblazon scholarship brochures at these institutions. The relationship between these top institutions in AI and computer science is deep, financial, intellectual, and intimate.

In 2024, the top ten undergraduate programs in computer science in the United States were (1) Stanford, (2) MIT, (3) Carnegie Mellon, (4) UC Berkeley, (5) University of Illinois Urbana-Champaign, (6) Princeton, (7) Cornell University, (8) Georgia Tech, (9) University of Washington, and (10) California Institute of Technology.[18] The highest ranking Historically Black College and University (HBCU) in computer science, Howard University, was not ranked within the top one hundred. For the most part, there is a certain permanence to these top-tier institutions. For the last fifty years, there has been very little movement among them. That suggests another embedded institutional pattern that exerts its own force on the challenges of participation. Like anything else, access to the top is limited. This is especially the case in advanced computing, due in part to the cost of computing power and the inertia of research development and integration with the leading tech companies. They have a special flux of intellectual and financial resources that is self-sustaining and keeps them structurally above the rest of the university world. When MIT announced its Schwarzman College of Computing in 2018, for example, it was a result of a $1.1 billion investment from the institution, catalyzed by a single $350 million gift.[19] By comparison, Howard University's total endowment in 2019 was $693 million—and Howard has the largest endowment of all the HBCUs in the country. MIT, at the time, had an endowment of more than $17 billion—and they are roughly the same size schools, both serving approximately twelve thousand to fourteen thousand students. MIT has an endowment-to-student ratio of $1.4 million per student, while Howard's ratio is approximately $48,000 per student.

There is no novelty to the idea that the most elite institutions have a gravitational pull. They draw the most talented, imaginative, and gifted academics from around the world, and they can afford them. It has been a hallmark of American higher education since the end of World War II. Elite research universities in the United States are among the best in the entire world. It is particularly true in advanced computing and engineering. Evidence of that is found in the international diversity of the graduate programs where the cutting-edge research takes place. The very best students from around the world make up the majority of graduate students in computing and engineering at the top US universities, and increasingly, the faculty are from all corners of the world, particularly China and India. In 2024, nearly 60 percent of all the PhDs awarded in computer science, computer engineering, and information science from the top ten programs in the United States went to international students.[20] And the trend, prior to the administration's efforts to curtail it, was continuing. That same year, nearly 70 percent of enrolled students in PhD programs in those fields at those schools were international students.[21] It is one of the crown jewels of American technical innovation and research.

In the twenty-five years between 2000 and 2024, approximately ninety thousand students earned undergraduate degrees in computer science from these top ten schools.[22] This was also during the meteoric rise of

Big Tech—the ascension of the FAANGs: Facebook, Amazon, Apple, Netflix, and Google. The combined valuation of these companies in 2024 was more than $9 trillion. Over the same time period, approximately 2,200 Black students earned undergraduate computer science degrees from those institutions. In twenty-five years, only 2,200 Black students earned undergraduate CS degrees from the top ten institutions in America combined. That is less than 90 per year from ten schools. Over the first quarter of this century, 2 percent of computer scientists from the most elite institutions in the country were Black. To give some perspective, the high school I attended in Brooklyn, Midwood High School, currently has more than four thousand students enrolled. At Google there are approximately sixty-five thousand tech workers. If they hired every one of the twenty-five years' worth of Black CS graduates from the top ten institutions, they would make up approximately 3 percent of their tech workforce. In 2023, at the very highest level of academic achievement in CS, there were six Black students who earned PhDs in computer science from the top ten institutions combined. Six out of 562—1 percent. Beyond the top ten, of the approximately three thousand PhDs in CS granted in the United States in 2023 from all institutions, seventy-one of them went to Black people—2 percent. In a time of surging research in AI and quantum computing and the transformation of information and computing capacity, the likes of which we've never seen, seventy-one Black people earned PhDs in computing in the entire country.

Eileen's question of whether Google should buy or build a university revealed a level of power I hadn't considered. Setting aside the zero percent chance that an organization like Google would consider that in today's openly anti-Black season, the relationship between the exception and the pattern remains. The flip side of the patterns of Black students' achievement in algebra and the constraints on the path to the elite quarters of the AI and STEM world is the lack of exposure that high-achieving white and Asian people have to Black people. Here, too, the human element behind the numbers matters. The anchors to history that weigh down individual Black achievement at the highest levels have a corollary among white people. The likelihood that the highest-achieving white people in the tech enterprise know many, if any, Black people is small, and that lack of interaction is significant given the questions at hand. Their lack of exposure is significant because it limits their view of the

American experience. In this regard, the core arguments of the benefits of diversity are true.

The confluence of disparate worldviews, of different American experiences, is in fact necessary to weave together the whole American fabric. The interconnectedness of the community we are all members of is strengthened by our proximity to and understanding of each other. Among this population of rarefied white and Asian technologists, that exposure is more than a romantic notion of steps toward a more perfect union. It grounds their technical decision-making and calibrates it against their understanding of both the brutality and beauty of the American story. The assumption of the virtuous white American—and the belief that knowing better yields doing better—is undermined by friendship with and connection to Black people who know better than to believe in that American fable. Our lives depend on disbelief. Connection to Black people helps white people understand a version of the American story that they don't have access to. Applied to the massive artificially intelligent systems now under construction, this level of human interconnectedness is more important than ever. It informs white people's consciousness. It provides for a richer connection to a broader community and hopefully a greater sense of responsibility and openness to the potential risks hiding behind the noise of the benefits. It helps provide anchor to the considerations behind Eileen's question.

Their isolation is merely the flip side of our isolation. It is simple logic. The Big Tech corporate diversity season of highlighting exceptional Black people tells an interesting story beyond the marketing of false progress. Most major American cities are racially segregated.[23] That is certainly true of New York, where I'm from, and Atlanta, where I live. The segregation of their public schools reflects the racial segregation of the cities. The grand claims of diversity in both cities are true, but segregation requires diversity to be meaningful. In fact, segregation is more poignant the more diverse a city is. In 2023, the population in Brooklyn was 36 percent white people and approximately 26 percent Black people. That is also merely a data point in the passing migratory story of white and Black people, as is the case in several major urban centers. In 2000, the population of Brooklyn was 35 percent white and 34 percent Black. The population of Black people has dropped nearly 10 percent in almost twenty-five years. The out-migration of Black people is happening fast. Despite Brooklyn being more than a quarter Black people, there are hardly any Black people in Park Slope, for example, and there are virtually no white people in East Flatbush. More than 90 percent of the people in East Flatbush are Black and more than 70 percent of the people in Park Slope are white.[24] Both are neighborhoods in Brooklyn. Atlanta is one of the most segregated cities in the country too, and its population is nearly half Black and half white. I'll come to the matter of Asian and Latino people in this matrix later. Given how pervasive racial segregation is in America, the paucity of Black students on the most elite pathways to the frontier of the STEM enterprise means that most white people on those very paths have hardly any interaction with Black people.

If Stuyvesant High School is an example, the problem is clear. The 70 percent of white people who live in Park Slope live alongside only 6 percent of Black people. The elementary school in the heart of the neighborhood, PS 321, has approximately 1,300 students. Two percent of the students are Black, and 72 percent of them are white.[25] It is not only possible but probable that most white students who live in that neighborhood and attend that school don't have any Black friends or know any Black people directly. If the highest achieving of them move through a similar middle school experience and end up at Stuyvesant, the likelihood continues. If they attend Stuyvesant, where a freshman class consists of seven or eight Black students among eight hundred or nine hundred students, it is equally likely that most of the white students there will neither know nor befriend any of those handful of Black students. Notwithstanding the American instinct for racial tribalism, the actual numbers make it difficult, regardless. Most of the white students will never have a class with one of the Black students, thus taking away the primary means of meeting and making friends. There aren't enough Black students to spread out into the clubs or sports, so the opportunities to interact are limited. It is a probabilistic result of the sheer numbers.

If, from there, one of those white students from Park Slope who graduates from Stuyvesant decides to study computing at any of the top CS undergraduate programs in the country, they will encounter the same problem. Some years at my alma mater, UC Berkeley, there aren't any Black students in computer science at all. Only 4 percent of students in the whole university are Black, so there are equally small opportunities for white students studying CS at UC Berkeley to meet Black students even outside of their major course of study. It stands to reason that the pathways to elite, selective higher ed institutions are very similar across the country and are reflective of the resegregation of American education and the continuation of residential segregation. Most white students who attend these elite institutions will likely only know other white or Asian students. That will have been the case in all their Advanced Placement computer science courses, or their AP Calculus BC or AP Physics C courses, or on their Regeneron (formerly Siemens Westinghouse) Science Competition team, or their First Robotics Team, or their coding camps, and ultimately in the neighborhoods in which they live. In sum, their whole segregated educational lives. This is particularly true in major urban centers. White students who live in exclusive suburbs and exurbs may never even have met Black people.

So, when the student from Park Slope, or any similar neighborhood with its majority white and Asian public schools, goes to Stuyvesant or any of its peer schools across the country and ends up at one of the top universities, they will have had little or no direct exposure to Black people. Certainly not Black people who are their intellectual peers. That hurts them. It narrows their view of the American story and limits their ear for the breadth of the American voice. In public schools in New York, the janitors and food service staff may well all be Black, and most of the Metropolitan Transportation Authority (MTA) train conductors and bus drivers who get them to and from school will be, but not their peers. Even in Brooklyn, where Black people make up more than a quarter of the population, and in Atlanta, where they make up nearly 50 percent, the challenge is the same. Even in the schools that have both Black and white students, their experiences are notoriously divergent, with the white students in the most advanced courses—courses into which very few Black students gain access. This particular subset of white people who are among the highest achieving in primary and secondary school, and then specialize in computing at the highest levels, have virtually no interaction with Black people. It is the uninterrogated consequence of the patterns that trap Black people on the outskirts of the most advanced sectors of the STEM enterprise—the narrowness of white people's understanding of America.

Then, once they leave a school like UC Berkeley, where the percentage of Black students in CS is sometimes zero, or Carnegie Mellon, where it hovers between 2 and 3 percent, they move into a professional role at one of the Big Tech companies. Needless to say, the challenges there are exactly the same. They could only be, given that the most elite computing programs have been the primary feeder for the Big Tech companies for years. So, this rarefied group of white people who, along with their Asian colleagues, constitute the pinnacle of the American technical brain trust, have virtually no exposure to Black people. Of course there will be individuals with different experiences, but the demographic reality of this rarefied tech world suggests this is the norm—the other side of the pattern.

There is a kind of surreal imbalance here. The Black people they are exposed to are exceptional titans. They are the most highly educated and accomplished of their generation, with the hopes and dreams of entire communities on their shoulders. These Black people are the kind of people most people never meet. They are among the country's most brilliant technologists and have defied America's expectations of them and broken the pattern. In the case of Dr. Charles Isbell, for example, one of the nation's leading scholars in artificial intelligence, the current chancellor of the University of Illinois Urbana-Champaign and former dean of the College of Computing at Georgia Tech, he is the first Black person in MIT's 164-year history to get a PhD in computer science (1998). When they meet Dr. Marian Croak, a vice president of engineering at Google and the patent holder for Voice over Internet Protocol (VoIP), they are meeting the first Black woman to earn a PhD in social psychology and quantitative analysis from the University of Southern California (1982) and a graduate of only the fourth class at Princeton (1977) to allow women. Even just the mid-level Black software engineer is likely of the league of the six or seven Black students admitted per year to each of the nation's best public high schools and the 2 or 3 percent of Black students at any of the top institutions in the country. These Black people are giants, made of a kind of steel forged in American fire. This class of white people's exposure is, in many ways, limited to Black titans. All they know are exceptions. They never meet the pattern, other than to be served by them.

That dynamic offers some insight into why, during the diversity decade in Big Tech, the exceptional Black person was so celebrated and fawned over. The pattern of Black underachievement is the dominant public narrative. It is born of the American story that says that Black people are of second-order intellect. It permeates the air in this country. Black parents know that one of their principal duties is to inoculate our children from this American scourge. When our positions are incommensurate with that American expectation, the instinct of many white Americans is to say that we achieved by way of diversity and affirmative action, not by merit. What we have and have achieved was given, not earned. We are "diversity hires." As the president has said, we ought to be relegated to "Black jobs." Having seldom experienced Black people as peers, or even as friends, and certainly not as superiors in the tech space, they are understandably taken aback—not sure what to make of the incongruence. Hence, in the season when diversity became a social cudgel that the Big Tech firms couldn't ignore for fear of public backlash, the response was to highlight the Black people who were close. As a result, it crudely became high-end modeling season for exceptional Black people. Those Black students trapped in schools and circumstances circumscribed by the pattern were an interest of distant altruism. They may well have been refugees in some far-off African land. They were not the exceptional Black people who were close. There was scarcely any real expectation that students from beleaguered schools with inadequate academic backgrounds locked off in segregated neighborhoods with Black-sounding names could somehow, through superficial programs of tech altruism, cross the Rubicon and end up with real tech jobs in Big Tech companies and sit face-to-face as peers. And to the extent that those expectations were real, they were often held by Black people themselves in these companies and relegated to Diversityland—a kind of sincere but unserious world of performative social insurance given the power of the companies in which they sat.

In the end, and not unexpectedly, the approach of Big Tech to this challenge of the relationship between the exception and the pattern was to highlight the exceptions as indications of progress and ignore the patterns that made the exceptions exceptional in the first place. During the era of programmatic intervention for equity in computing, I participated in a meeting of Google leaders in education about the impact of the educational partnership work we were engaged in. All of the work was under the umbrella of expanding computing capacity—partly in the company's self-interest in developing more software engineers for the workforce, and partly as an act of civic responsibility from a corporation with near-matchless power. There was a component of the discussion that was deeply academic about the metrics of success. Is a measure of progress in equity in computer science education the increasing proportion of CS undergraduate degree holders who are Black? Is a more appropriate measure the number of students who enroll or indicate interest in postsecondary computing education? Or is it merely the number of students who, after having gone through some Google-sponsored program, indicate a greater sense of self-efficacy in the computing world? It was a range of considerations that varied between our effects on faculty as the unit of analysis to high school teachers and students themselves. In the end, the edict was that the top Key Performance Indicator, the KPI, was the number of "wins." For those who do not spend time in the corporate world, a KPI is a success metric. It is how an organization prioritizes its goals. They serve as the metrics against which progress is measured. Ultimately, they reveal the problem a group is organized to solve. Given that, what is a "win"? After grueling debates about measuring educational progress and the efficacy of partnerships, a win was the number of positive mentions of Google's work in the press and on social media. That was the top KPI: brand and reputation. Yes, there were others more germane to the core issue of contributing to equity in computing education, but none of those were first. The top KPI is the most revealing of the problem you're trying to solve. If positive mentions in the press and on social media is the top KPI, then the problem you're trying to solve is one of good social standing and brand reputation. In Kahlil Gibran's *The Prophet*, there is a section on giving. The rich man said to the prophet to speak to us on giving. Among other things, the prophet says, "There are those who give little of the much which they have—and they give it for recognition and their hidden desire makes their gifts unwholesome."[26]

It is easy to make the case that the responsibility of Big Tech companies does not include breaking the patterns that entrap so many young Black people and lock them out of the front of the STEM enterprise. One of the challenges, however, is that breaking the pattern is what they said they would do. For years, Big Tech companies proclaimed that equity in computing was a core value and that diversifying the tech workforce was one of their goals. The only way to do that is to break the pattern—hence Eileen's question of whether to build a university or buy one. While her suggestion, if realized, would not have affected the hundreds of thousands of Black students languishing and perishing in algebra, it was bold and, in the narrow world of advanced computing, would make a structural difference. It was the consideration of a powerful agent to at least chip away at the pattern in a substantive way. Notwithstanding that momentary consideration, the corporate response is always the same—bottom line, profit, shareholder accountability, core business, revenue generation, not our lane, we're not trusted actors in that space. Yes. We are so entrenched in our socialization of accepted norms that any pushback against that basic argument is regarded as naïve. "We will not be able to do any of this work if the company does not exist" is how the counterpoint is framed. There are organizations whose core mission it is to address those issues. All of those things are true, but they lack the broader vision of civic responsibility, imagination, and courage. They also lack the full assessment of the responsibility of power in a time when norms have been upended. The threats associated with the permanence of the patterns that entrap Black people are no longer merely a lack of access to the tech frontier. In a time of the convergence of AI and a surge of unbridled racism, the risk is of a descent into digital slavery for a people who have been there before. And Big Tech is building the tools to enable that to happen. They are building the equipment for modern slavery and not considering their responsibility to the modern slave.

CHAPTER 4

DIVERSITY DEFANGED

Conviction comes with cost. You can only measure that cost when the core values upon which that conviction rests run into opposition. It is easy to believe in freedom when you're free, and no one is threatening to take that freedom. Equally, it is easy to declare allyship and solidarity and support when all of those stand you in good favor. Doing so when it comes at personal risk is another matter. "If I stop to help this man, what will happen to me?" That question is the calculation of the cost of conviction. In the brochure version of modern American cities pocked with utopic life, Instagram images of everyone's dream version of themselves, and the rise of Big Tech's gilded class, it is sometimes hard to see the basic calculation of conviction that is constantly taking place. People are calculating, institutions are calculating. How much am I willing to risk for someone else? How much cost am I willing to wager for an ideal that I profess to hold dear? How much individual risk am I willing to take in defense of the public good? Who is worth defending in America, and at what cost? Or, said another way, who is worth sacrificing, and for what gain? How much resistance to the caste structure will I wage?

In his 1992 book *Faces at the Bottom of the Well: The Permanence of Racism in America*, Derrick Bell, one of the nation's preeminent legal scholars on school desegregation and the first tenured Black faculty member at Harvard Law School, lays out the idea of America's sacrificial people elegantly in his short story "The Space Traders."[1] In it he weaves a tale in which aliens from outer space come down to the United States in 2000 to make an offer. The small delegation of aliens emerged from a fleet of fantastically large box-like ships, bigger than aircraft carriers and harkening back to our imagined images of Noah's ark. The fact that they arrived in ships on the East Coast of the United States, between Cape Cod in the North and North Carolina in the South, was a foreshadowing of their haunting offer. It turns out the ships were filled with three resources the United States would need to return to full glory. One was a supply of gold, enough to restore the backing of the US dollar and completely erase the national debt. The second was an otherworldly chemical that could reverse the pollution of the nation's soils, rivers, and waterways that had been destroyed by unchecked industrialization. The third was a renewable source of clean energy to permanently relieve the toxic burden of burning fossil fuels. In sum, they were offering a restoration of the country—a reset to the days when it was under the balanced stewardship of Native Americans. In exchange for the treasures, they wanted the United States government to give them one thing—all the Black people. They wanted every man, woman, and child who was classified as Black in America to be sent to them, to board the otherworldly galleons to be taken back with them to their distant home in outer space. They didn't say what would happen to them there, good or bad, and they said no force would be applied to the deal; it was entirely up to the government of the United States to accept or decline the deal as offered.

In the story, white people looked at the ships, and they conjured up wholesome, majestic images of the Mayflower that arrived on the American coast in 1620 to start the great national experiment. Black people, however, looked at the ships with dread as they unlocked horror embedded in the marrow of their bones—horror associated with the *White Lion*, the first slave ship to arrive in America in 1619, the year before the Mayflower, to start the great national experiment. In the story, the Space Traders give the United States sixteen days to come to a decision. The sixteenth day coincided with Martin Luther King Day that year. The story goes on to outline the haranguing of the government and the American people, the arguments for and against from different segments—the business community, the legal community, and general popular opinion. In an eerie foreshadowing of the senior-most federal cabinet officials today, those in the story are all white, with a token person of color who is willing to scorn the heritage and struggle of his own people with his obsequious deference to an ideology that despises him. In the story, that cabinet member is Black, and Black people refer to him as Uncle Tom. The white members of the cabinet almost immediately conclude that Black people are worth sacrificing. They swirl around an unspoken belief that they all share, that white America doesn't want them anyway, so why continue to pretend? Black people are just a problem. However, they offer different reasons. Some suggest that Black people should consider it an ultimate act of patriotism, a kind of martyrdom in service of the land of the free and the home of the brave—an act of love for a country that will love them for the first time once they're gone. Others suggest that getting rid of Black people will be a benefit to the country beyond the alien treasures offered. It would lower the welfare rolls, ease the burden of prisons, rid the nation of violent crime, free up the health care system to care for the legitimately sick, and unchain public education from its perennial underperformance. Most poignantly, members of Congress are flooded with calls, particularly from Conservative districts, that their constituents are pleading with them to accept the offer. They want their elected officials to accept the offer, round up the Blacks, send them to outer space, and make America great again. They see it as a kind of final solution. At the cabinet level, Uncle Tom realizes, finally, he is in fact Black. His obsequiousness will grant him no hiding place in the end.

As we might expect, the story concludes that America votes to send Black people away on the last observance of Martin Luther King Day. Black people, a kind of voiceless citizenry in the negotiation of their fate, knew that the end was near. To the white liberal voices of protestation, the story suggests, "Black people know your plans for legislation, litigation, and protest cannot prevail against the tradition of sacrificing black rights. Indeed, your efforts will simply add a veneer of face-saving uncertainty to a debate whose outcome is not only predictable, but inevitable." The story leans on the real tradition of skepticism born of the Black experience in America.

Black people have etched into their memory Article 1, Section 2 of the Constitution of the United States, which states that each Black person is only to be counted as three-fifths of a person.[2] They have etched into their memory the 1857 US Supreme Court decision in *Dred Scott v. Sanford* that ruled that Black people cannot sue for freedom because they are neither citizens nor people; they are property.[3] They have etched into their memory that while in 1865 the Thirteenth Amendment of the Constitution abolished slavery and involuntary servitude, it left an exception for prison labor and convict leasing, which led to a century of slavery by another name.[4] They have etched into their memory the 1896 US Supreme Court decision in *Plessy v. Ferguson*, which allowed for legalized racial segregation and laid the legal groundwork for nearly seventy years of an apartheid state in America under Jim Crow laws.[5] In more recent times, Black people watched as the *Brown v. Board of Education* decision in 1954 made school segregation by race illegal, but white communities chose to close public schools instead and send their children to private schools so they wouldn't have to sit next to Black children.[6] Black people remember that while the Fourteenth Amendment guarantees equal protection under the law, they watched the 1973 Supreme Court decision in *San Antonio Independent School District v. Rodriguez* allow for inequitable school funding.[7] All students must have access to public schools, yes, under the equal protection clause, but they don't have to be equally funded. Black people have etched into their memory that the Voting Rights Act of 1965 was designed to finally give them the right to vote and stop the terror and repression of Southern states' disenfranchisement of that most fundamental civil right.[8] And they remember, fewer than fifty years later, when the act was gutted in 2013 by the Supreme Court decision in *Shelby v. Holder* striking down Section 4(b), which monitors election changes in states with a history of blocking Black people from voting—allowing those states to pick up right where they left off.[9] Even today, Black people watch on TV and social media as innumerable Black people are chased, strangled, shot, tazed, stomped, and bludgeoned to death by police and emboldened white vigilantes.

The evidentiary basis for Black people being a sacrificial people in America is irrefutable. In the story, Black people conclude that the hopeful opposition of liberal white people is smothered by Black life experience. The government accepts the offer from the aliens. The insurrection of an indomitable people is violently smothered by the military and national guard that the president was all too happy to unleash. The government rounds up the nearly forty million Black people, approximately 13 percent of the American population, and sends them off in alien ships. Their lives don't, in fact, matter. They leave just as they arrived 381 years before. And the American whites quietly said, "Amen."

The idea that Black people are a forsaken American people is central to the current assault on Diversity, Equity, and Inclusion. The tawdry debate about DEI cheapens its deeper roots in the American narrative. Narratives have ancestors too. In this case, the long and tortured internal struggle within the American body of how to reconcile with its original sin. Black people are not only the substrate of the sin of slavery white Americans and Europeans committed for multiple centuries, but we are also constant contemporary reminders of that sin. Black people's pledge to never forget has the corollary that white people must always remember. We were both involved in the experience of this American original sin. It was a ghastly but intimate affair. While white people would like to forget for fear of confronting their past, Black people can't forget because of the implications for our future. Hence, the willingness of the US government in "The Space Traders" to allow for a path to forget by getting rid of the source of remembrance: Black people.

Absolution of sin and guilt is a far easier proposition when the object of sin and the source of the guilt—if guilt exists—is rendered invisible. It makes it much easier to forget and use that posture of having forgotten as proxy for having been forgiven. It is much easier to say, "Those were my ancestors; I had nothing to do with it," when not confronted with the obvious racial consequences born of those ancestral sins. The mere presence of Black people in America is a constant source of that internal tension built into the American narrative. The common refrain among the white people attacking the teaching of complete American history, for example, is precisely that. The attack on DEI is a thinly veiled attack on this very issue. It is the heart of the matter. It is an attempt to solve the problem of the recollection of ancestry by erasing learning and interrogation of the origins and practice of the nation's original sin. It is claiming, "Those were my ancestors, but I had nothing to do with it." Blocking students' opportunity to learn about the full American story, which includes barbarism and shocking levels of inhumanity, is an attempt to sanitize it and make the dark segments disappear. It runs farther away from the core issue of having to ask Black people for forgiveness. God can absolve the sinner of their sins, but forgiveness is a worldly affair. Erasing the history, or as in the Space Traders story, erasing Black people themselves, removes the requirement of forgiveness—a prerequisite for a functioning multiracial, multiethnic society in the United States. But the idea of American white people memorializing and asking American Black people for forgiveness is so antithetical to who and what America is, it is nearly un-American to even think it.

Despite that troubling truth, the approach to sanitize history underestimates the power of the narrative's ancestors. Ignoring it, ceasing to teach it, doesn't make it disappear. You don't in a family say, "Those were my ancestors; they have nothing to with me." Rather, the reality is that I am me *because* of my ancestors. If not for them, who am I? In that framing of interconnectedness, you cannot be accused or held to account for the individual sins of a past relative; of course not. But the attitudes and sentiments of a family transcend generations. Ironically, it is the very bedrock of the Make America Great Again thesis that is driving the assault on DEI. They reference American traditions, the intangible chords of values and ideology and ways of being in the world that connect one generation to the next and constitute the cultural fabric that made America great in the first place. They're calling on the ancestry of a particular narrative to reemerge. They're conjuring it, calling it up, from the blood and soil of the nation's past. They offer the same question: *But for the American ancestors, who is America?* They see themselves as originalists. The challenge, of course, is about which ancestors they're calling. What is the original story they'd like to be told? Which traditions and spirits are they conjuring? Are they calling the white man astride his horse, armed with a bullwhip and salt, and all he represents? Or are they calling the Black man with a tattooed back, facing endless rows of cotton, and all he represents? The challenge comes when the white Americans who attempt to call on the founding documents and fathers of the nation as objective progenitors of American truth overlook the fact that the founding fathers were astride the horse. They wrote the original documents when it was illegal for Black people to write. In *The Souls of Black Folk*, W. E. B. Du Bois raised the question that the descendant of the overseer asks of the descendant of the overseen: "How does it feel to be a problem?"[10] It feels like knowing you are of a people that your nation would forsake.

The experience of being a problem in society is the experience of being the sacrificial people. It is knowing where you are in the rank ordering of people: at the bottom. It is the experience of having your presence being tolerated as opposed to assumed. In the current tussle over DEI, the signals abound of who the sacrificial people are. With regard to Black people in the American story, there has never been any ambiguity. Du Bois goes on to lay out one of the most famous paragraphs describing the contours of a sacrificial people.

> After the Egyptian and Indian, the Greek and Roman, the Teuton and Mongolian, the Negro is a sort of seventh son, born with a veil, and gifted with second-sight in this American world,—a world which yields him no true self-consciousness, but only lets him see himself through the revelation of the other world. It is a peculiar sensation, this double-consciousness, this sense of always looking at one's self through the eyes of others, of measuring one's soul by the tape of a world that looks on in amused contempt and pity. One ever feels his twoness,—an American, a Negro; two souls, two thoughts, two unreconciled strivings; two warring ideals in one dark body, whose dogged strength alone keeps it from being torn asunder.

The question that remains unasked and certainly not answered is the counter question from the overseen to the overseer: "How does it feel to have caused the problem?" The resistance to that question, the absolute refusal to accept ancestral responsibility and exorcise ghosts, is fueling the inertia of this current anti-DEI assault. It is not a specious debate about whether preference or merit is the bedrock of American achievement. Accomplishment and achievement have always been some combination of favor, preference, merit, and good fortune. It obviously works differently for Black and white people, but nothing is pure in that regard, and we all know it. The resistance is rooted in the issue of whether the country will take responsibility for the bifurcated racial patterns that exist everywhere, in everything, in every dimension of American life. There is absolutely no escaping it. The entire basis of Black exceptionalism is based on the universality of American racial patterns.

The resistance to DEI is a resistance to accepting the coexistent burden of racism and responsibility. It isn't a concern about merit or divisiveness. Clearly, the resistance to it is pulling the sides of the American population farther apart, so divisiveness isn't really a concern. Those resisting are willing to tolerate divisiveness when the issue that causes it is their own. The recollection of ancestry in white American slavery and its descendant narrative is the throughline of racism coursing through the nation's veins. It is the ancestral connection to the overseer astride his horse. The refusal to accept responsibility for that narrative and the shocking implications that have guided and shaped the behavior and personality of the country ever since is the bedrock of the anti-DEI movement. It is a refusal to acknowledge racism and accept responsibility for the national soul that has given it safe harbor through the ages. There is a level of introspection and contrition that the purveyors of this anti-DEI assault are fighting in a powerful cultural and political tantrum. Understandably, it is surely difficult for the monster to reflect on what makes him monstrous. James Baldwin, in *No Name in the Street*, offers an elegant insight into that reflection: "In the private chambers of the soul, the guilty party is identified, and the accusing finger, there, is not legend, but consequence, not fantasy, but the truth. People pay for what they do, and, still more, for what they have allowed themselves to become. And they pay for it very simply: by the lives they lead."[11]

The life the United States is leading now is clearly a life of divisiveness and discord. It is unclear who all this chaos benefits and who is happy as a result. The peace and calm of a settled heart and clear national conscience is elusive. We are constantly at war with our own demons. We are paying for what the American ancestors astride the horse have allowed us to become. And we are paying for it with the lives we lead—angst, anxiety, rage, fear, despair, and discontent. Where is the safe harbor? The American tribes enjoying the ascension of their ideology and relishing in their refusal to take responsibility for the ancestors of their narrative can only be concerned about themselves. It can only be a kind of supremacist narcissism. The absence of charity and kindness, the unwillingness to give, the prevalence of meanness and assumed criminality all anchor to the view of the tattooed man from the perch on the horse. That Black man and all that he represents will neither break nor go away, and this is a constant source of anger. It is a constant pick at the scab on a guilty heart.

The man on the horse would rather build two separate school systems than allow him entrance. Would deny him both a job and the safety net he needs because he can't get one. Would privatize prisons, create laws to imprison him, and profit from his imprisonment. He would scream and yell that Black students have stolen the seats of deserving white students when only three out of every hundred students enrolled in elite computing in America are Black, when only six out of more than five hundred candidates who earn PhDs in computing in a year are Black. He doesn't want him to gain an inch. He doesn't want him to acquire anything at all, certainly not power, and is willing to scorn and forsake any attempt at national harmony if the people with the tattooed backs refuse to submit to being a sacrificial people. He would call in the military when Black people merely declare that their lives matter. When they look around and see the chaos, the fear, the loathing, anguish, and acrimony that they've sown, what possible joy can they derive from what amounts to collective sorrow for us both? To the extent that they derive joy from the suffering of others, it is evidence of the continuity of the ancestral narrative.

It was the bullwhip that rendered the tattoo. What was in the heart of the man astride the horse that enabled him to draw satisfaction from cracking the whip and blasting open the flesh of the man he sees as thing? There is a psychosis in that. Schadenfreude is gentle. Deriving joy or satisfaction from that manner of barbarism is a kind of sadism that helps us understand the foundation of the original ancestral American narrative. It also helps us understand the ferocity of the resistance to looking inward for absolution. The instinct to lash out, to further demonize, diminish, and demolish, to erase efforts to learn and interrogate that past is almost understandable. It likely engenders fear of what they might experience when they look upon themselves in the ancestral mirror and, as a reflection of a past self, see the white man astride the horse drooling venom and the supportive community that enabled him.

In the season of American lynching, thousands of Black people were lynched by marauding white mobs.[12] The rampant violence and viciousness are hard to consider as acts of human beings, but the normalization of it, the community acceptance, is another level of depravity entirely. White communities would congregate with families and children, have picnics, and gleefully take pictures with disemboweled, castrated Black people hanging from their necks in the background. What is the message to the child in the example of the parents in that scene? What was the attitude of the photographer, and what did he say—"One, two, three, smile"? What was the family narrative born of that tradition? The point here is not to be gratuitous in the articulation of white American sadism but to offer shape to the fuel driving the intensity of resistance to DEI and the emotional place from which it may be coming. Interrogation of that ancestral past must surely be deeply traumatic—pitiable, even. Their argument that the study of this past will make white kids feel bad about themselves is acknowledgment of the barbarism. Hence, the effort to smother it and to blame the descendants of the tattoo for the moral distress of the descendants of the bullwhip. The path to absolution, however, runs through truth, not around it.

That resistance and obstinate refusal to bend in front of obvious moral turpitude is central to the problem at hand. It may well be that the proportion of American white people who are driven by and applaud this resistance to introspection, reflection, and absolution is not the majority. It is hard to know. In 2024, the President of the United States was explicit in his declaration to embrace moral depravity, racial scorn, vengeance, and social hierarchy. He, a white convicted felon, did not hide it. Going down the line of *what-ifs* is not intellectually enlightening, but the idea of a Black convicted felon, a man for whom the US prison complex was designed, running for and winning the presidency of the United States isn't the stuff of make-believe. It is so inconceivable to be nothing at all. In this case, there was no Trojan horse. He said exactly how he felt, and his colleagues shared their plans explicitly. There was no ambiguity. Yet 56 percent of all the white people who voted in the election voted for him. They overlooked his criminality, his low moral character, and his tradition of scorn and meanness and supported him anyway. That is approximately sixty-two million white people in support of unfettered racial animus, embedded with the ancestral American narrative of Black people as a sacrificial people.[13]

It will be left to future historians to look back and ask what happened to America during this dark regressive season of her life. Surely, even now, the arguments about various issues—the economy, so-called kitchen-table issues, and others will conflate and confound. There will be a lot of erudite pontifications. But, as the ideology pertains to the issue of diversity and all that it represents, there are some very simple through lines when considered from the vantage of the descendant of the tattoo. The narrative that is descended from the tattoo offers clarity of sight. As Du Bois said more than a century ago, the Negro is "gifted with a second-sight." That sight offers an uncomplicated view of the historical pivot, because we have seen it before. I am neither swayed nor bothered by the inevitable arguments about the number of Black men who sided in favor of the ideology of the bullwhip. There were Black men who, when given the chance, sat astride the horse too. The issue is the refusal of those supportive of the bullwhip ideology to kneel and seek absolution. That is the issue. That is what is driving the resistance to DEI.

That refusal, and what they are willing to destroy in defense of it, is the outline of the challenge we now face as a country, together. It is the much deeper issue cloaked by the DEI argument. It harkens back to the reality that while the Civil War was fought over slavery, and the Union ultimately prevailed, it was the bloodiest war Americans ever fought. In four years nearly 650,000 people were killed—and not by advanced weapons of mass killing or bombs dropped from fighter jets. It was hand-to-hand combat. Close, intimate killing. The kind that is born of deep-seated defense of an ideology. The Southern white Confederates were willing to forsake the very existence of the country in defense of slavery. They were willing to slaughter, destroy, maim, and die to keep Black people as things enslaved. Their descendants suggest that they were fighting to defend states' rights or the sustainability of their cotton-driven economy. But at the center of either of those causes were Black people. Four million enslaved Black people who, if abandoned to states' rights, would have remained abandoned to slavery. The cotton-driven economy was built on centuries of their stolen labor. Historical attempts to obfuscate that basic truth are the things we've seen before. When future generations of the descendants of the man astride the horse look back at this time, they will use the same arguments. It had nothing to do with race; it was the economy. It was inflation. It was government fraud. It was immigration. It was crime. Again, we'll surely hear, as we are now, that it was states' rights. This is the pattern that is revealed with the gift of second-sight. What we see is the brutal tantrum of a restive people unwilling to confront the nation's past. We see a willingness to rise and smash progress gained, as was the case at the end of the first Reconstruction and here now at the end of the second.

The DEI argument—as high stakes as it has become because of implications to law, educational standards, and funding both in the private and public sectors—remains at its core about the refusal to take responsibility for past wrongs. That refusal to accept responsibility for the sin of slavery and its torturous aftermath is what is at hand. It trivializes the issue to suggest that slavery was so long ago that even mentioning it is the flailing complaint of a whining people searching for something to blame for our shortcomings. The mere fact that we are still here, standing and making the case, is testimony to fortitude, perseverance, and belief in a better nation. We believe in the nation that has offered precious little to anchor that belief in. And yet, here we are facing the slogans: DEI is radical terrorist ideology, teaching WOKE ideology is an insidious indoctrination, DEI is the epitome of divisiveness, DEI is an abomination in the American classroom and cancerous to the country. All of this angst is not only the tantrum in the face of confronting a dark past; it is also resistance to recognizing the parallel humanity of Black people. That, ultimately, is the absolute core of the American sin and the source of this ongoing battle. While the DEI moniker has come to represent all manner of Americanisms, it was the anti-Black loathing that was the engine of slavery and the founding of the country. That hatred for Black people is the cornerstone of American racism and a founding doctrine of this nation. The fight over DEI is not only an unwillingness to acknowledge that the American doctrine has always had Black people as a sacrificial people, but also a willingness to make sure that they continue to be.

I served a three-year term as a program officer at the NSF in the Directorate for Computer and Information Sciences and Engineering (CISE). It is the directorate in the agency that supports basic research in computational sciences. The research in computational science that has given rise to artificial intelligence, quantum computing, and the frontier of American computing was and continues to be funded by the NSF. It is a significant part of the engine that powers American technological leadership in the world. Not only is the scientific enterprise in computing reliant on the research funded by the NSF; it is also central to the structure of the US national system of innovation. At risk of sounding wistfully romantic and hyperbolic, basic research is the bedrock of human understanding. Despite this country's innumerable ailments, the research infrastructure in the United States is one of the global gems this nation has resoundingly contributed to the progress and betterment of the human condition. President Obama used to say that while we are shackled to the demons of our past in many ways, there is no better time in human history to be a human being than here in the early part of the twenty-first century. That is thanks in part to the international reach and standards of the United States research infrastructure.

Because being a program officer of the National Science Foundation is a federal position, you must take the federal oath. For people who have never worked for the government or been a public servant, it is hard to convey the moment. Even for someone like me, highly conscious and critical of what the American government has done to Black people, it is a towering honor. The complications of a double consciousness notwithstanding, the ability to serve, to contribute to the public good, to help advance human understanding and improve the human condition is an extraordinary feeling. The military is lionized in the narrative of national service for the defense of the country. People clap for them at airports. But in the world of advancing human understanding and unlocking the mysteries of science, serving the US scientific enterprise is an honor second to none. For a moment, as you take the oath, hope smashes cynicism. It is hard to convey what that feels like. In the oath, you say out loud:

> I, Kamau Bobb, do solemnly swear that I will support and defend the Constitution of the United States against all enemies, foreign and domestic; that I will bear true faith and allegiance to the same; that I take this obligation freely, without any mental reservation or purpose of evasion; and that I will well and faithfully discharge the duties of the office on which I am about to enter.

After the swearing in, you have to attend a small group orientation. It took place in the US Capitol building. The civil servant who conducted mine was a short white woman who sounded like she was from Yonkers. I don't remember her name, but in my memory, she seemed like a Laverne. Part of her official role was conducting the orientation for freshman members of Congress and new federal employees. She was old and seemed to have been working in the Capitol since a time when slaves were building it. She may have been about 4'10" tall, and she was bent. Her voice was raspy and loud, and she smelled like coffee and cigarettes. She said she'd been doing these orientations for nearly fifty years. She cursed when she talked. She gave us some details about the functioning of the government and the relationship between the various branches and how the checks and balances work. But she was clearly edgy or angry. She seemed frustrated with us despite having just met us. She said that, over the years, newly elected officials and new officers know less and less about the US government. She said their egos are getting bigger and bigger, and their knowledge is getting smaller and smaller. She said, "A lot of the new ones don't know shit about this country. The best they've got is *Schoolhouse*

Rock." I was taken aback, but it was hilarious. Her style didn't match the occasion or the location, a hallowed room within the temple of American democracy. But then she turned on a dime to a tone of solemnity. She said we were being entrusted with the American people's faith. That the thing that makes this country work, despite all its challenges, is the American people's faith in the government, that the American government is supposed to work in the best interest of the American people—that it is the bedrock of the American democracy. She paused. Then she said that has not always been the case, its history is checkered, but it is a sacred trust that we are signing onto, and it is our job to make the government deserving of the faith entrusted in it. We are not only taking the American people's faith; we are taking their money and asking them to trust us with it. She finished by saying, "In every decision you make, ask: Is it worthy of the faith entrusted in you by the American people?" The orientation felt like a scene from a movie, as if there should have been some patriotic soundtrack playing in the background in the voice of James Earl Jones.

I had never directly confronted my personal relationship to the founding documents and what they mean and what fidelity I have to them, or they to me. In my world, these documents inscribed a permanent degradation of Black people. The fullness of me as a person, and the people from whom I come, had to be edited in. The Three-Fifths Compromise had to be undone by the Fourteenth Amendment, but the language of both is still there. Taking an oath like that, which says that I will bear faith and allegiance to the Constitution, forces, at least for me, some pause. Does the Constitution and what it stands for bear faith and allegiance to me?

I left the orientation and went to the National Archives to see the original Constitution. It sits in the Rotunda for the Charters of Freedom along with the Declaration of Independence and the Bill of Rights. I wanted to at least see it. The original Emancipation Proclamation, signed by President Abraham Lincoln in 1863, is also in the archives, and I wanted to see that too. I wanted to see those two documents, the Constitution and the Emancipation Proclamation—the existence of the latter confirming the shortcomings of the former. When you go into the Rotunda, the lights are dim, and the humidity is controlled. There are armed guards standing alongside the case where the Constitution sits. The solemnity of the room is staggering. It is a shrine to the founding documents of the country. You cannot escape the gravity or significance. It is the kind of space that forces deference and respect. In another section is the Emancipation Proclamation, where, in the original handwriting, President Lincoln declared:

> That on the first day of January, in the year of our Lord one thousand eight hundred and sixty-three, all persons held as slaves within any State or designated part of a State . . . shall be then, thenceforward, and forever free; and the Executive Government of the United States, including the military and naval authority thereof, will recognize and maintain the freedom of such persons, and will do no act or acts to repress such persons, or any of them, in any efforts they may make for their actual freedom.

Constitutional scholars and historians have belabored the significance of these documents, about their utility, for ages—some idealistically, some cynically. I am neither a constitutional scholar nor a historian. I am a descendant of the tattoo trying to put some historical context to the current assault on DEI and what it presumably means. In the tumult about such issues, when Black people refer to slavery, the attackers of DEI and the silent exasperation of many people suggest that they feel it is ancient history, so long ago that it has lost relevance today. Slavery lasted for 244 years in the United States and was the subject of the bloodiest war on American soil in history. The Civil War was finally won in 1865, more than a century and a half ago. The end of slavery—the longest, darkest, and most grim committal of American sin, the bedrock of the nation—was only formally acknowledged for the first time in 2021. The acknowledgment in 2021 came at nearly the end of the first quarter of the twenty-first century for something that happened in the latter half of the nineteenth century. It was first acknowledged 156 years later when Juneteenth was designated a federal holiday—when the Juneteenth National Independence Day Act made it law. Despite the protestations that it was too long ago to matter, that it is ancient history, the reality is that it took the country more than 150 years to even symbolically acknowledge the end of its most shameful era and that Black people were free at last from bondage. That acknowledgment happened only four years before the nation's 250th birthday. Clearly, it is still relevant and resistive. The timing of the federal acknowledgment of Juneteenth pulls it into this century and makes it relevant and belies all the critiques to the contrary. The past is present. It is not lost on me that the institution of slavery existed nearly as long as the United States has been in existence—244 years and 250 years, respectively. The celebration and memorializing that take place on Independence Day to acknowledge the founding of the country is testament to its historic significance and all that it implies about the course we have been on as a country together. The reluctance and grudging offering of that same recognition for the freedom of Black people is equal testimony to the value the nation puts on that milestone. It says in the Emancipation Proclamation that the Executive Government of the United States will recognize the freedom of all persons formerly held as slaves. At least symbolically, that didn't happen until I was forty-nine years old.

After the George Floyd killing in Minneapolis, the country and world went into revolt and had a massive racial reckoning. The uproar over that murder catapulted the Black Lives Matter (BLM) movement, which began after the killing of Trayvon Martin in 2012, to the front of national consciousness. As a symbolic gesture to the simple statement that Black lives do, in fact, matter in America, the saying began to appear on stickers and murals across the country. It was evidence that the historical narrative of the nation needed a reminder. In a nation where Black people's presence is, in the main, tolerated and not assumed, it requires stating explicitly that our lives matter. It is not the refutation of any other life; rather, it is an affirmation of the life of Black people. The fact that the nation needs a reminder is evidence of the patterns of violence and disregard that have plagued the country forever. It is another link to the descendant narrative of the man astride the horse. In his version of American history, Black lives did not matter. They could be disposed of like chickens or hogs—animals with which their value was equated. Per the *Dred Scott* decision enshrined in the record of the United States Supreme Court, Black people were merely property.

In Washington, DC, like in so many other cities around the country, the Black Lives Matter mural appeared. In DC, it was officially sanctioned by Mayor Muriel Bowser in 2020 and emblazed on Sixteenth Street near the White House. The short stretch was renamed Black Lives Matter Way and the vicinity declared Black Lives Matter Plaza. 2020 was one year before the historic recognition of the end of slavery and the establishment of Juneteenth as a national holiday. It is clear that Juneteenth was declared a national holiday in large part because of the Black Lives Matter movement. It was a triumphant, hard-fought, albeit symbolic, moment acknowledged because of the national upheaval and the towering courage of the young leaders of the BLM movement. It is something for which the nation, in a more honorable stance, would simply say, "Thank you."

In 2025, under command and threat from the federal government, the District of Columbia was forced to erase the mural, just five years after it was established. What does that really mean? What is the symbolic experience? It wasn't merely painted over. The administration ordered that the street upon which the words "Black Lives Matter" were emblazoned be destroyed. Heavy equipment, jackhammers, and demolition tools were used to smash the concrete and cut the steel chords that held it together. It was a physical destruction and desecration of the symbol declaring that Black life, that my life, matters in America. That literal and symbolic destruction of the declaration that Black Lives Matter was in Washington, DC, the capital of the United States, the mountain top of the federal government. What more symbolic beacon could there be indicating that Black lives don't matter than to have the federal government of the United States destroy the moniker saying that they do? Of course, the adherents to the thesis of the man astride the horse offer the spurious argument that Black Lives Matter is a divisive slogan, and they smile at its destruction. But divisive to whom? To whom is it a divisive idea that I suggest that my life matters, other than to someone who thinks that it doesn't? On the contrary, so valuable are white lives to the American narrative that innumerable Black people have been killed just for looking at white people—Emmett Till being perhaps the most glaring example of that American trait. There are no pictures of white people hanging by their necks from lamp posts in the background as Black families pose with their children celebrating such a macabre scene. There aren't any in American history, not a single one. There isn't now, nor has there ever been, any systemic threat to white lives coming from Black people.

The declaration that Black lives matter has nothing to do with the value of white lives. Rather, it has to do with the killing instinct of white people toward Black people, rooted in the relationship between the bullwhip and the tattoo. The continued refusal to repent of that murderous instinct, to kneel and seek absolution, is evidenced here in 2025 in the destruction of the Black Lives Matter mural. The segment of the white American population that is holding onto the narrative of the bullwhip will simply not let it go. Referring to that reluctance to surrender to better angels and submit to contrition and absolution as being anti-DEI is a euphemism. It is a mask for a much more sordid American psychosis and a renewed threat to the safety and security of Black people and anyone else on the list of others. Instead of contemplating reflection and national healing and harmony, we're experiencing more flailing, and destructive petulance—a doubling down and a resurgence of a real threat of violence. The Emancipation Proclamation declared that the "Executive Government of the United States, including the military and naval authority thereof, will recognize and maintain the freedom of such persons, and will do no act or acts to repress such persons, or any of them, in any efforts they may make for their actual freedom." Yet we stood and watched as the executive branch of the United States government destroyed the sign declaring that Black lives matter.

These are the powerful undercurrents of the American narrative flowing below the debate over diversity, equity, and inclusion. Given that, the result of the collision between Big Tech and DEI yields outcomes that are predictable, if not inevitable. If Black lives don't matter, surely their employment in the high-tech sector is hardly of national concern. The tech titans stood proudly alongside the president after he publicly declared that Black people eat dogs and that Black jobs are the bottom rung jobs, rendering Black people in competition with undocumented workers—who he has dehumanized and criminalized. Yet there they stood. What else ought we expect? The decade of work in support of DEI in the tech world, from 2014 to 2024, yielded some progress, yes. But it always felt like the grudging commitment of reluctant companies with no particular ambition for success. In some ways, it felt as if the problem was manufactured in their world and wasn't, in fact, a real problem. The problem statement was never quite clear. Hence, success or failure was always vague. It was never even fully clear what success meant.

One of the core pillars of the argument was underrepresentation. The underrepresentation of Black people, and others, was a problem. The determination of that as a problem, however, was asymmetric. It was a problem for Black people to be excluded from the most elite technical sphere and all the power, influence, and income that represents. Equally, it is a problem for Black people that they are not involved in shaping the technical future—not only the development of AI and its assorted tools but also the guardrails to help protect vulnerable communities. While that makes sense from the point of view of the excluded, it is less clear how that constitutes a problem for the companies. What difference to them does it make whether 2 or 4 percent of their tech workforce is Black? Hardly any. Therefore, from the companies' point of view, representation goals were a byproduct of the manufactured problem of underrepresentation.

The underrepresentation of Black people in the tech sector, however, is merely a secondary effect of the deeper American problem, which feels intractable and scary to name out loud, so it is left unspoken. But even sticking with the representation challenge, what does success look like there? Parity was bandied about. Parity with what? Ought there be an equal percentage of Black people in tech roles in Big Tech companies as there are in society? Fourteen percent of the population of the United States is Black. Ought that be the representational goal? Eleven percent of college graduates in the United States are Black. Ought that be the goal? At each level of filtration, the normative question remains. What is the right representational level? Two percent of students graduating with degrees in computing from the top ten institutions are Black, is that the right level? In the research quadrant of advanced computing, 1 percent of PhDs in computer science earned at the top institutions in the United States per year are earned by Black people. Ought only 1 percent of people in computing research be Black? It is an impossible question to answer. It's normative and depends on one's point of view. To the Black person, strong arguments stand for any of the parity measures.

The most poignant argument derives from access to the power associated with the elite tech sector at this stage of human development. That is a basic argument of justice, not merely representation. We have to be involved in the authorship of the future of global commerce, culture, and interconnectedness as we turn to the second half of the twenty-first century. We will either be authors of the story or merely subjects in it. The twentieth century was spent fighting to assume rights as full citizens. The twenty-first needs to be an exercise of the power derived from those hard-won rights. From the point of view of leaders in the tech sector, however, it could very well be that none of it matters, because the more fundamental issue of whether Black people are there at all isn't a problem.

To them, it is not a personal problem. It is a socially manufactured problem that they, perhaps with some degree of variability, believed in but bear no consequence if solved or not. The asymmetry in the way the problem is articulated contributes to the weakness of the DEI effort. It makes it easier to see why the leaders were so quick to abandon their already weak commitment to the issue in the first place.

While the president's destruction of Black Lives Matter Way is the most symbolic indicator of the value of Black people, the normative questions about representation lead to a plot twist in the story of Big Tech's commitment to DEI values. The signal that Black lives don't matter is an indigenous American philosophy and comes as no surprise. In Derrick Bell's *Faces at the Bottom of the Well*, he articulates that reality elegantly:

> Black people are the magical faces at the bottom of society's well. Even the poorest whites, those who must live their lives only a few levels above, gain their self-esteem by gazing down on us. Surely, they must know that their deliverance depends on letting down their ropes. Only by working together is escape possible. Over time, many reach out, but most simply watch, mesmerized into maintaining their unspoken commitment to keeping us where we are, at whatever cost to them or to us.[14]

Its corollary, however, is that white lives do matter. The higher value on white lives is the basis for the white supremacist ideology that has played out all over the world across so many centuries. But, more specifically, if we translate that into what is happening in the world of elite technical education, it poses a looming conundrum and potential threat. What happens when the underrepresentation is white people, not Black people? The underrepresentation of Black students in elite STEM education and highly selective American higher education in general is so commonplace and assumed that they have a name—underrepresented minorities (URMs). Black students have been reduced to URMs. It sounds like a disease. Institutions speak of the percentage of URMs they have. How gross. But so deeply entrenched is the nomenclature that minority is used even when they are not in the minority—hence the term, *majority minority*. On its face, the language doesn't even make sense. However, will we begin to describe elite American higher education in STEM fields as *minority majority* as Asian students start to outnumber white students? How do the normative pillars shift? It will add to one of the tenets of the brewing populist storm that somehow white people are being discriminated against. Discriminated against by whom remains a mystery. The noise and angst about Black people in the representational framework at the elite levels has always been that they are getting things they don't deserve. That our presence is an affront to meritocracy. It is the cornerstone of the anti-affirmative action argument. Less qualified Black people are taking the seats of white and Asian students who are more qualified. Quite apart from whether there is any merit in that claim, it is easy in America to offer that whatever Black people do is second-rate. It is the basis behind the president's claim that low-tier jobs are "Black jobs." Those are jobs commensurate with the lower intellectual capacity he assumes we have. It is the same sentiment Justice Scalia offered a mere decade ago, when he said that Black students ought to be on a slower track where they are more able to keep pace. It is a continuation of a long American narrative, which is why, when the nation heard it, some people were offended, but the idea wasn't new to anyone—slave labor, sharecropper labor, convict labor, janitors, kitchen staff, caretakers, app drivers, and shoppers. There are such things as Black jobs in their American framing. It is all consistent. In that framing, the underrepresentation of Black students in elite American STEM education and in the advanced tech ecosystem is just how it is and ought to be. But what happens when white students can't compete? What happens when, measured against their sacrosanct merit, white students aren't good enough? That will be a challenge to the American perception of the natural order of things.

The ease with which Big Tech companies, and many in the tech sector more broadly, were willing to drop their commitments to DEI is in part because of the symbolic equivalence of DEI and Black people. Of course, the leaders of the tech sector are not just white; there are many Asians as well. The truth is that the gravity that pulls Black people to the bottom of the American value-ranking of people is not exclusive to white people. It is part of the national narrative. At this stage of the American arc, referencing and articulating the reality and structure of white supremacy is hardly enlightening, notwithstanding the resistance to its premise. While there is massive resistance to the mention of white supremacy, its corollary, Black inferiority, garners no such clamoring. But the two need each other. If there is a group on top, so must there be a group on the bottom.

The reality of that hierarchy in this society, and frankly in most societies where Black and white people coexist, has been articulated ad nauseam by now. White supremacy and Black inferiority have been abusive American lovers since the nation's inception. They cannot divorce, however, unless they're both destroyed. But the idea of whiteness and its assumed power is admittedly tough to vanquish. Toni Morrison asked the question most elegantly, "If I take your race away, and there you are, all strung out. And all you got is your little self. And what is that? What are you without racism? Are you any good? Are you still strong? Are you still smart? You still like yourself? . . . If you can only be tall because somebody is on their knees, then you have a serious problem. And my feeling is, white people have a very, very serious problem."[15] Turning your back on an entitled top ranking is tied to the challenge of reflection and walking toward absolution. The rigidity of that supremacy affects everyone.

The significance of that ranking and associated power for Asians and other immigrants is not trivial. The idea of white adjacency is anchored to white supremacy. It exists because of white supremacy. White adjacency is the instinct to be close to white. To be, look, and sound as close to white as possible. It is an instinct to take on the mores of the presumed successful white American. It is a form of cultural adaptation. In its simplest form, if you can't be on top, you want to be as close as possible. It is a logical deduction. Upon arriving in the United States and calibrating against the social order, it is quickly clear to any new arrival who is on top and who is on the bottom. Then, in an attempt not to start at the bottom, one of the first instincts is to assert, "At least I am not Black." That instinct includes the adoption of American scorn toward Black people, almost as a means of fitting in and self-protection.

Subscribing to the tenet that the bottom tier of American society is a place reserved for Black people is, in some ways, how people know they're American. It is an adoption of the American caste system. Over the ages of American immigration, everyone has faced some version of denigration and American derision—the Irishman, the Italian, the Jew, the Pole—but in the end, they could all find shelter in being white. They could shed the parts of their identity that misaligned with American norms and hide in being white as a form of camouflage. It serves as hiding place, shield, and safe harbor. For everyone else, not being Black is the next best kind of insurance against not being on the bottom. Hence, white adjacency. If you cannot be white, then you can be as close as possible. Adjacency, in the context of the sweeping retreat from DEI efforts, means alignment with the American caste structure that renders Black people on the bottom. Given that, the extent to which the leadership of the Big Tech world is multiethnic is not protection for the fundamental tenets of diversity. If cultural norms can be predictive of individual instincts, there is no safe harbor. The adoption of the idea that there is a social ranking and that Black people in this country are on the bottom is not hard to see.

Of course, DEI has come to mean and represent many things that constitute the beauty and richness of American diversity. The whole effort, however, was born of Black people's exclusion and the push to right that perceived wrong. The occurrence of such a tepid defense of DEI in the first face of resistance is in part because of the hierarchy that we're so accustomed to and the willingness of so many to adopt it. It may even be that we don't even see it in the way we can't see the air we breathe. In Isabel Wilkerson's famous book *Caste*, she outlines what this reordering is up against.

> Caste is the granting or withholding of respect, status, honor, attention, privileges, resources, benefit of the doubt, and human kindness to someone on the basis of their perceived rank or standing in the hierarchy. Caste pushes back against an African-American woman who, without humor or apology, takes a seat at the head of the table speaking Russian. It prefers an Asian-American man to put his technical expertise at the service of the company but not aspire to be the CEO. Yet it sees as logical a sixteen-year-old white teenager serving as store manager over employees from the subordinate caste three times his age. Caste is insidious and therefore powerful because it is not hatred; it is not necessarily personal. It is the worn grooves of comforting routines and unthinking expectations, patterns of a social order that have been in place for so long that it looks like the natural order of things.[16]

Pushing against or resisting a caste hierarchy is a cultural commitment. Accepting the lower-tier placement of Black people is easy in America. But what happens when white people can't play? When their assumed perch as American Brahmins seated at the origin of the Cartesian plane—where relevance is measured by distance from them—is upended, then what? It is happening in the tech world. While the noise has been about Black people, where the changes have been minuscule, huge demographic shifts are taking place between whites and Asians. Even the reference against which Black representation is measured—whites and Asians—is swiftly losing significance. They are not a monolithic pair. Asians are quickly supplanting whites in the elite tech sphere. Over the same diversity decade of 2014–2024, at Google, for example, the demographic shift between those two groups has been mind-boggling. In 2014, 62 percent of the Google tech workforce was white, and 35 percent was Asian. Ten years later, 42 percent of the Google tech workforce was white, and 51 percent of the workforce was Asian.[17] The percentage of white representation dropped 20 percent in ten years. That is an astonishing shift. Their drop in representation was matched nearly one-to-one with the increase in Asian tech Googlers. That shift occurred quietly while all the noise and attention were focused on Black and Latino people. Kudos and salutations were being administered, and slogans of "we're making progress, but there is more work to do" were chanted, because over ten years the company went from two Black tech workers in one hundred, to four. That is what minuscule and hardly noticeable looks like. If there are four of you in a group of one hundred people, you may not even see each other. All the while, for white tech Googlers, they went from sixty-two of every hundred, to forty-two. That is a dramatic demographic shift from nearly two-thirds of the tech workforce to less than half in a decade. It cannot be clearer than at this stage of the Fourth Industrial Revolution that the captains of the tech industry are among the most powerful people on the planet.[18] For white people to have their seats at the table of power taken over by Asians is a challenge to caste rank. And challenges to that rank order are historically not well-received. The backlash, as Black people know, can be vicious.

Over the last twenty-five years, the demographic shifts at the most elite higher education institutions in computing have been remarkable as well. Here, too, the noise about Black people and agonizing fights over affirmative action have been deafening. Slamming the door to Black access to elite institutions has been taking place apace, but the percentage of degrees earned by Black undergrads in computing from those top institutions dropped from 3 percent to 2 percent. All the hysterics about Black theft of white seats is merely a form of the continued effort to keep

Black people in their place at the bottom of the well. All the while, the shift in the white and Asian block has gotten far less attention, because they are called out as one group—white and Asian. But they are not. In 2000, more than half, 53 percent, of all undergraduate degrees in computer science and information systems from the top ten computing programs in the country went to white students—*majority majority*—as it ought to be. Crudely, at that point, all was right in the world, the structure was intact, the order well kept—white people reigned supreme. However, in 2023, a mere twenty-four years later, less than a quarter, 21 percent, of those degrees went to white students. In four-year undergraduate cycles, twenty-three years is approximately six complete generations of undergraduates. In that time, the percentage of white students receiving those degrees dropped more than 30 percent. At the very time when the concentration of power in the elite tech sector is irrefutable, white students are losing their grip. They're slipping.

By contrast, over the same period, the percentage of Asian students has skyrocketed. In 2000, 27 percent of the degrees from the top ten computing institutions went to Asian students. In 2023, that increased to 42 percent. Those are Asian students who graduated from high school in the United States; Asians represent only 6 percent of the population of the United States. That is a staggering increase and overrepresentation. While 21 percent of the computing degrees went to white students, 75 percent of the US population is white. Also, over those years, the percentage of international students increased from 9 percent in 2000, to 19 percent in 2023.[19] Nearly as many international students graduated with degrees from the top ten institutions as did American white students. The vast majority of the international students are Asian too. They come predominantly from China, India, Taiwan, and South Korea. Including both domestic and international students, the percentage of Asian students earning undergraduate degrees from the top institutions is approaching 60 percent. Perhaps the data point that is the most foreboding is that, in 2023, there were more Asian American women who graduated from the top ten institutions in computing than there were American white men. When the throne of the white man is threatened, trouble is ever near. When there are tears in the caste, it fights back. It resists. As evidenced by the chants from the tiki-torched far right in recent years, white men do not like to be replaced.

For people in the academy who pay attention to such things, it raises questions. Asian students are not a monolith. Where are they from? Are these Chinese students, Southeast Asian students, Indian students? What socioeconomic strata do they come from? Are they the crazy rich Asians of modern imagination? Are they first-generation immigrants, or have they been in the United States for generations? Are they descendants of the Chinese who built the transcontinental railroad? Are they descendants of the Japanese Americans who were forced into internment camps in WWII? Are these students multilingual in addition to being multicultural? Do they see themselves as Asian American or as international students? All of those things matter to people who care about national and international systems of education and the sanctity and value of individual students and their identities. They matter to people who care about who these students are and how best to support and educate them. It also raises questions about the hubs of the best STEM education in the world and what their strengths and shortcomings are. What are the cultural attributes that engender highly mature capacity in numeracy, logic, and creativity—the bedrock of advanced STEM education? What are the curricula and pedagogical traditions in primary and secondary education that are the most effective? Are there replicable approaches that may help bolster the US STEM education ecosystem? But, in the world of DEI assault and the crudeness of the place from which it comes, those details don't matter. What does matter is that Asian students are replacing white students, and in the elite STEM education sphere, it is a zero-sum game. Colleges have fixed capacity. If the same crudeness of the anti-Black sentiment holds, whether a student is from South Korea, Singapore, Taiwan, China, or Queens is irrelevant. They aren't white. That is the important distinction. And that trend is irreversible.

The top tier of STEM education in the United States is dominated by Asian students. The data on that front are undeniable. Even looking backward for a moment at the SAT, however fraught that exam may be, the patterns are clear. If we use it in the way that many do, as some God-sent standard of merit, white students can no longer compete. According to the Brookings Institution, the correlation of race and performance on the SAT bears that out.[20] In 2020, before all the COVID disruption, 35 percent of Asian students who took the SAT scored better than 700 on the math section. Sixty-two percent of them scored better than 600. By contrast, a mere 9 percent of white students scored better than 700 on the math section, and only 31 percent scored better than 600. At the most elite American institutions of higher education, where applications are surging and standardized tests and supercharged GPAs are coming back into favor, white students are fading to the second tier.

In the degenerate language of American stereotypes, it is common to say that "Asians are good at math." However, Asian students on average outperformed white students on the reading section too. Eighty-three percent of Asian students exceeded the College Board's benchmark for college-level reading, as compared to 79 percent of white students. If we use the same crude logic of merit that is hammered on Black students as a cudgel against our intellectual capacity, the same conclusion must be drawn. Maybe white students, according to Justice Scalia's line of thinking, need to go to slower-track schools where they would be more comfortable and able to keep pace. Perhaps forgo college altogether and get certificates and leave advanced computing for Asian students. So goes the Scalia argument; so comfortable was he diminishing Black people. Is the country comfortable coming to the same conclusion for white students? Is the country comfortable relegating white students, who are typically the ones associated with elite education—wealthy and upper-middle-class white students—to a lower tier of . . . anything? It is antithetical to who and what America is. That class of white people may very well join the populists in violent defense of caste. They may already have. The replacement of white students with Asian students at the most elite institutions in the United States is a disruption of our caste ordering of people.

There is no reason to think that this trend will reverse. The dramatic shifts started to accelerate in earnest between 2010 and 2020. That decade was a time when the country was emerging from the economic crisis of 2008 and the coincident accension of Big Tech. Fifty-one percent of the degrees from the top computing institutions went to white students in 2010. By 2020, in just ten years, it plummeted to 28 percent. If that approximate rate of decline continues, and there is no indication that it will not, the percentage of white students earning degrees in computing from the top institutions will be down to 5 percent by 2030. There will be nearly no white students left in elite higher education computing in the United States. They will be competing for the 2 to 4 percent of seats that Black students occupy. The outcome of that game of musical chairs is preordained.

What will the attackers of DEI do when white people can't compete for seats in the advanced US technical enterprise? I concede this whole line of argument is crude and reductionist, but it is in keeping with the environment in which we now sit. To address the issue, we clearly need a removal of the polite language that obfuscates. The insulting and degenerate attacks on Black people from the president and the adherents to his way of thinking have set the stage. It is not lost on me that the surge of anti-Chinese sentiment coming from the government and the populist adherents is coincident with the surge of Asian students at America's most elite institutions. Neither is it lost that these institutions are currently under withering assault. There has to be growing resentment among the class of white Americans who have for so long felt entitled to seats in those most hallowed institutions and entitled to their legacy. I cannot be convinced otherwise. White students who ride the crest of their wealth and privilege and access to everything enriching work hard, do all the right things in their rarefied space, but are still, in the end, outrun by Asian students. That has to hurt.

The growing vitriol toward China and Chinese people has some roots in the technology sphere. The claims of their theft of US patented technologies and suspicions about the use of apps like TikTok to infiltrate Americans' privacy for nefarious use by the Chinese government are signals. While the broader issue of Chinese technology in competition with US tech interests is a far more complicated techno-economic issue, the language is a signal of the growing sentiment. The Chinese technologist is a danger to American hierarchy. In the world of contemporary crudeness, just as the immigrants crossing the southern border have been transformed into Mexican and Venezuelan narco-terrorists regardless of where they may be from and what their human narrative is, the Asian can be reduced to an odious Chinese tech spy. In that framing, it is logical that Asian students from abroad seeking to attend elite higher ed institutions in computing in the United States simply have their student visa applications declined. In keeping with the demographic trends, they are already on pace to outnumber and significantly replace white students at these institutions by 2030. That is a threat. The revocation of visas is one lever, another is that much of the federal funding that supports the graduate research in their fields is cut off, leaving many international Asian students with no way to support themselves—a strategy of starving them out. In the first five months of this administration, through May of 2024, funding from the National Science Foundation was reduced by 51 percent. In computer science, the reduction was more than 30 percent.[21]

There is precedent for when caste moves to defend itself. It fights back. The very first immigration law in the United States to ban a specific group of people's entry into the country was to ban Chinese people. The Chinese Exclusion Act of 1882 put a federal ban specifically on Chinese immigration for a decade, from 1882 to 1892.[22] The similarities are haunting. The popular claims at the time were that Chinese people were stealing jobs from deserving white Americans and that they were a dirty people sullying the clean American fabric, albeit 1882 was less than twenty years after the end of slavery. While there is not yet any public outcry over Chinese people stealing jobs from white people in this era, the data from Google is a proxy for what is happening at the frontier of the tech sector. Over the last decade, the percentage of Asian tech workers has skyrocketed while the percentage of white tech workers has plummeted. Similar to the demographics at the nation's most elite STEM education institutions, it is plausible for there to be hardly any white people left at the elite levels of the tech sector in the next decade or two. In an environment as racially fraught as is currently the case in the United States, that could very well be conceived of as a threat. It could be an ominous foreshadowing of a threat to "white jobs," to their supremacy in the world's most powerful sector. It is particularly ominous because this threat to the caste-based top ranking of white people in this critical technical arena is taking place within another replacement threat lurking in the background.

The United States will cease to be majority white this century.[23] After two and a half centuries, the country will no longer be a predominantly white nation. White people under eighteen years old are already minorities in several border states. They are the new minorities, upending the idea that the United States is a white Christian nation. This demographic shift is a definitional revolution. It is a foundational contradiction to the central claim of when America was great. Ostensibly, America was great when it was a white Christian nation—when Native Americans were vanquished, Chinese were excluded, Japanese jailed, Black people chained, and others struggling to survive under the pressure of caste. It was great when there was no contestation to that foundational definition—when the top seat on the caste system was white and unchallenged.

Now challenged, we are experiencing the rage and opposition to that idea. We are experiencing caste fighting back. The attempt to round up, kidnap, and deport 10 to 12 million predominantly Latino people, the fastest growing population in the United States, is evidence of that resistance. Between 2022 and 2023, the Latino population accounted for nearly 70 percent of the population growth of the whole country.[24] As a group, they are posing the largest threat to white dominance of the population. The Trump government's solution is to get rid of them—to round them up and snatch them off the streets, from their homes, from the nation's schools and churches, and send them to ghastly detention camps in border states and in other countries. Immigration and Customs Enforcement officials, agents of the federal government of the United States, in the mold of the Gestapo, are marauding Latino communities, kidnapping parents from their children, raiding workplaces, attacking people in their cars, and making a public spectacle of degradation and humiliation of an entire group of people.[25] It has an old feel, like gatherings at the lynching tree, like Kristallnacht. In the Caribbean, with the assistance of the US military, they are murdering Latino people on the high seas. Their bodies, burned and drowned by American hellfire, are drifting ashore in Trinidad.[26]

It is a stunning return to racist and genocidal instincts that we swore to never forget. It is looking into history and bringing back the trains. Here, too, what clearer signal can the government send of its wrath ignited by the perceived threat to the American caste structure? There are both historical and contemporary examples of how far American caste, in the body of the federal government, will go to defend the ranking of people and ensure that white people stay on top. It can be brutal and tyrannical. The history of examples is too long to list. To think that Asian students and people will somehow be immune from this gathering storm of vengeance and violent defense of the rank order is naïve. Just as Uncle Tom in the Space Traders story discovered to his dismay that, in the end, his obsequiousness couldn't save him from being Black, here, too, neither white adjacency nor model-minority status will calm the wrath of caste raging to defend itself.

The signals are creeping in. In addition to the potential theft of jobs, there is the growing suspicion about the character of Chinese students, faculty, and researchers. What is their intent? Are they agents of the Chinese Communist Party? Are they spies? Are they coming to study advanced AI and quantum computing technologies only to return with them to China to compete against US tech companies? We are in an era when students are being arrested on college campuses and suspicion of the intentions of both faculty and students is on the rise. Historical precedents offer a warning. Chinese and Japanese people cast as spies, as dangerous subversive operatives for foreign states, has happened before. The Japanese have obviously been subject to the wrath of caste too. President Roosevelt's executive order in 1942 enabled the broad roundup of Japanese Americans for forced removal into internment camps. In the relatively unsullied first quarter of this century, we have largely been lulled into thinking that such things cannot happen in the United States. But we now have seen hundreds of Latino men from the lower tiers of the caste structure whisked off American streets, heads shaven, shackled and sent to prison camps with no other charge than of being lower caste. We have seen legal American permanent residents and students at prestigious US universities arrested from their homes and off the streets and disappeared. Even the language is odd when referring to the United States. For my generation, people are "disappeared" in distant, brutal dictatorships—South America, Southeast Asia, Eastern Europe, Africa—not in New York. We don't even use the word in that context in this country. People aren't disappeared; they disappear.

When caste fights to defend itself, it isn't concerned about the details of the presumed enemy; it is concerned with the preservation of self. It is a fight to maintain the position of American white people at the top of the rank order. Asians of all nationalities and ethnicities can be swept up in the vengeance of caste's defensive fury, and the vitriol against China is ratcheting up, signaling the gathering storm. This transference of the target helps us understand how diversity has been defanged and what the resistance is really about. When DEI is seen to mean supporting Black people in the Big Tech world, there is no willingness to pay the cost of having conviction. There never was any substantive conviction to begin with. The marketing of core values is not the same thing as having them. Caste just whispered its discontent, and within days all the commitments in Big Tech came crashing down—people fired, programs decommissioned, budgets slashed, titles removed, reports discontinued, declarations of DEI goals erased. We must adhere to federal executive orders. We must do as we're told and forsake those faces at the bottom of the well.

Now that the threat of underrepresentation, or no representation at all, is pointing at white people, the narrative must shift. Because DEI is fraught and painted Black, another strategy must be born. There are hints of one strategy taking shape. One way to address the underrepresentation of white students in elite computing spaces is to remove the Asian students who are causing the underrepresentation—stripping of visas, denial of entry, logistic and legal intimidation, cascading chilling effects. Since the argument of merit demonstrably doesn't work with Asian students, more crafty means of resistance will have to arise. In some circles, hijacking the term DEI to make it mean geographic diversity is a play. That could mean that schools have to admit students with some fixed representation from certain geographic areas, presumably regions with low percentages of Asian, Black, and Latino students. It could be that the political leaning of a state is taken into consideration as a means of ensuring equal representation of conservatives. Columbia University has already been ordered to exercise comprehensive reform of its undergraduate admissions process. Of domestic students, its school of engineering is 18 percent white and 28 percent Asian.

The conviction associated with maintaining the caste ranking comes with a willingness to pay any cost. Given that, we should not underestimate the lengths to which the descendants of the man astride the horse will go to reframe and restructure the landscape of the frontier of the American tech enterprise. The defense of American caste is a ruthless affair. In the context of Black people, it is fueled by a refusal to come to terms with a barbarous past that requires absolution and healing. It comes with punishment for even questioning the rank order and our position at the bottom of the well. How dare we? In the context of Asian people, it has more to do with the defense of hierarchy. While Black people were never a threat to the substantive hierarchy in the tech sector, Asian people absolutely are. In many ways, they have already taken it over and reordered the social ranking. Hence, DEI as an arm of justice to support Black progress in Big Tech is done. It is thoroughly defanged for this era. Reframing the debate of underrepresentation now that it applies to white people is likely what the next chapter will be. Of course, the difference is that framing the effort from the white top of the caste structure is to frame the effort with fangs.

PART III

21ST CENTURY, SECOND QUARTER

CHAPTER 5

MANDELA AND ALTMAN

I saw Nelson Mandela in person in June 1990 in Brooklyn. It was five days before I graduated from Midwood High School, and I had just turned eighteen years old about two weeks before. I didn't meet him, but I saw him. By now, his story is a sacred chapter in world history—a story of justice, human dignity, nation building, and Black liberation. At the time, however, he had just been released from prison in February of that year and was on his first trip to the United States four months later. Apartheid was still in place. It would be another four years until the first free elections, which brought the African National Congress (ANC) to power and Mandela to the presidency of a reborn South Africa in 1994. At the time, what is now history was future. We didn't know. We just knew it was him: Nelson Mandela. On his first trip, his very first stop was in Brooklyn at Boys and Girls High School in Bedford-Stuyvesant. Before he went to Gracie Mansion, where the mayor of New York City resides, or the ticker-tape parade in Manhattan, or City Hall, before he went to Yankee Stadium to address all of New York, he came to Boys and Girls—our school, deep in Bed-Stuy. Even if you didn't go to Boys and Girls, its name rang out. Brooklyn people know: Fulton Street between Utica and Schenectady. In 1990, Bed-Stuy was a Black place. His first stop was to come see us. Black people. He came to hug us and let us know that it would be alright. He came to let us see him and hold his hand and tell us that he loved us too. He told us he knew we were also struggling. He came to thank us on behalf of Black people in South Africa for our support and prayers in their life-and-death struggle against the Afrikaners. To have

Nelson Mandela come to Brooklyn, come to us, and say thank you—the feeling is beyond the reach of words. He came to tell us that we are one people with a common struggle. He came to remind us what freedom looked like and what it meant to believe in justice. And I believed in him.

Boys and Girls was a special school. Not only is it the first public high school in Brooklyn—its history dating back to the late nineteenth century—it is a Black school in the heart of Bed-Stuy—a historic Black neighborhood—and it was home to the African Street Festival. The African Street Festival in Brooklyn is one of the largest and oldest summer festivals of its kind in the country. It is a complete celebration of Pan-African life that began in 1971. West Indians, Africans, Black Americans, and Black people from every corner of the African diaspora gather in a way that is only possible in New York. Every summer, hip-hop, dancehall, and Senegalese talking drums rang out, and roti, ribs, jollof rice, and injera were dished out. It was a Black love fest. Thousands of Black people converged, and for those few days we could lay our armor down and be free. For Black people, it was an annual pilgrimage. In the late 1980s, the festival was more than just a summertime breeze out; it was a sanctuary, a kind of refuge.

For Black people, the late '80s in New York was a walk through the valley of the shadow of death. It was the height of the crack nightmare. When crack addiction was killing Black people, the government launched a war against us—and the addiction led us to war against each other.[1] The violence and incarceration were jaw-dropping. Different from now, when opioids are killing white people and the government has declared a national health crisis and launched a healing campaign.[2] Not only were the police running wild against Black people, especially young Black men, the media was whipping the racial rancor into a frenzy.[3] The Central Park Five incident, the case where five young Black men were falsely accused of brutally raping a white woman, had just happened. Donald Trump put out full-page ads in local newspapers explicitly calling for them to be killed.[4] A mob of white men in Bensonhurst had just the year before attacked and shot Yusef Hawkins to death for being in their neighborhood. A different mob of marauding whites chased and savagely assaulted three Black teenagers in Howard Beach after their car broke down. One of them, Michael Griffith, while trying to escape, ran onto the Belt Parkway and was hit by a car and killed. Bernhard Goetz, a white vigilante ancestor of George Zimmerman—Trayvon Martin's killer—shot a group of Black teenagers on the train because he didn't like how they were behaving. The NYPD had broken into Eleanor Bumpurs's home and shot the elderly Black woman to death at close range with a 12-gauge shotgun, claiming she attacked them. For jumping a turnstile, the police arrested Michael Stewart, a famous Black graffiti artist, hog-tied him, and beat him to death in their custody. And Rudy Giuliani, in his failed bid for the mayorship of New York against our first Black mayor, the Honorable David Dinkins, unleashed a tirade of racist insults, innuendos, and provocations—not dissimilar to those of President Trump—that riled up the mob.[5] Black teenagers especially—we were catching hell from each other, from the marauding white mobs, and from the police. There was no safe harbor.

As such, Boys and Girls, and the grounds of the African Street Festival, were a kind of hallowed high point to rest and breathe and look above the nadir. It was here that Mandela came. Because I grew up in a Pan-Africanist household, I had a poster of Mandela on the wall in my room. It was of him when he was young, the last time anyone had seen him in those days, and it highlighted the end of his freedom gospel delivered at the Rivonia Trial, after which he was sent to prison for all those years.

> During my lifetime, I dedicated my life to this struggle of the African people. I have fought against white domination, and I have fought against Black domination. I have cherished the ideal of a democratic and free society in which all persons live together in harmony and with equal opportunities. It is an ideal which I hope to live for and to achieve. But if needs be, it is an ideal for which I am prepared to die.[6]

On the day, Boys and Girls was covered in a blanket of Black people. There were Black people everywhere. The grounds are surrounded by a fence. The fence was covered with people. The A train runs right by the school on an elevated line on Atlantic Avenue. There were people on the steel trusses perched for a glimpse. Every time a train stopped, from every door in every car, people flooded out. We were gathering. Sisters with gold Fulani earrings and cowrie shells. Egyptian musk was in the air. Incense.

Rastas. Brothers with High-Top Fades and Gumbys. Gold grills. It was an extraordinary experience.

I was in high school when I started to learn what was really happening in South Africa. My father made me read about the desperate struggle there. I had to learn about how the ANC tried to critique and protest the Apartheid laws that led to the forced removal and displacement of nearly 3.5 million Black South Africans and required that they carry passes to indicate permission to move about in their own country. But, at a point, the wanton viciousness and human depravity of the white Afrikaner government rendered the path of protest suicidal. They turned to guerrilla warfare. Nelson Mandela founded and led the paramilitary arm of the ANC, *Umkhonto we Sizwe*, Spear of the Nation. Their battle for freedom was at all costs. It was a life-and-death affair. It was of one accord with Malcolm X's message, "We declare our right on this earth . . . to be a human being, to be respected as a human being, to be given the rights of a human being in this society, on this earth, in this day, which we intend to bring into existence by any means necessary."[7]

When Mandela finally arrived in his motorcade, the crowd surged. When he walked onto the small stage he sucked the air out of our lungs. We gasped, breath suspended, and then released. People exploded into cheers. We'd never met this African hero. Most of us had never even heard his voice. For people my age, we'd never seen anyone like him. Malcolm X: killed. Martin Luther King Jr.: killed. Medgar Evers: killed. Fred Hampton: killed. We'd never met a Black superhero who stood in the center of the ring to fight toe-to-toe against the racist killing machine and lived to tell about it. At that point, Mandela had been in prison my entire life. It was a kind of rapture. A larger-than-life freedom fighter who connected the struggle of our people on the other side of the Atlantic—the other side of the Middle Passage—to us in Brooklyn, on Fulton Street. There he was in person. Maya Angelou says that you don't necessarily remember what people say, but you remember how they make you feel. I don't remember what he said. I hadn't noticed I had tears in my eyes. But I remember feeling free. And I loved him for it.

Freedom and justice were at the heart of that experience. We were an embattled people, a congregation of beleaguered believers in that core human instinct for freedom and demand for justice. If Mandela were to be Christlike, my faith in him would be based on his demonstrated commitment and willingness to die for my freedom. I believe in that. He doesn't need to make the case about the relationship between his struggle and the core principles of a democratic and free society. I understand it in my bones. I trust the argument. He didn't need to ask me to believe. My instincts rendered me a believer. We have to use the tools at our disposal, be they technical, social, violent, nonviolent, or political, to build and protect freedom and justice that are the cornerstone of a free and just society. That belief proposition is clear, and the yearning and struggles of people around the world to achieve that end are universal evidence of its clarity. While he said that he has fought against white domination and Black domination, he didn't have to. The larger goal of freedom and justice demands that both be expelled.

Now, in the time of AI, we are confronted with different apostles and a different creed—technological progress and power. Technological progress is threatening to wrest the altar from freedom and justice. AI has wrought a new era of empire. The consolidation of power and wealth associated with the few major tech companies vying for dominance of AI has created a new world reminiscent of the old, and its leaders are asking us to believe in them.[8] For those of us with roots in the colonies and former colonies of Europe and America, we know instinctively what that means from our end of the experience as the colonized. It means dominion and subjugation. It means resource extraction, theft of wages, and extortion of labor. It means existing in a kind of cultural fog that blurs our foresight and muddies our memory. It means being Black and living in a country named British Guiana, or French Guiana, or Dutch Guiana. Or being Zulu living in your own country and needing a pass to leave your home. It means existing in a hierarchy so deeply entrenched that even the idea of a Kenyan Britain, or an Algerian France, or a Congolese Belgium is so far-fetched that it doesn't even exist in imagination. The concentration of power and voracious ambition to win at all costs among the tech companies and their financiers at the top of the AI pyramid brings us back to a different time. While the circumstances are materially different, the features are hauntingly similar. Once upon a time, as the saying goes, "the white man came with the Bible, and the Black man had the gold. Centuries later, the white man has the gold, and the Black man has the

Bible." The captains of these companies are proselytizing. They are luring us into believing in the grandeur of technological progress. They are seeking to have us place our faith in them—in their technical wizardry and self-proclaimed prophetic witness. But where in that creed is justice? In this new gospel, according to the tech captains, technological progress is Lord of lords and King of kings. But from the world of the tattoo, justice is the rock upon which I stand. Technological progress is incidental to the march of scientific development. Justice is not. In the world of education, the distinction holds—technological progress and educational justice are not the same thing. Justice is not an assumptive outcome of progress. On the contrary, the kind of progress—whether it be fair or foul—depends on the explicit commitment to justice.

The investigative journalist Karen Hao outlines the religious zeal of the leaders in the AI world in her book *Empire of AI.* Among several far-reaching observations of the stunning suck of natural resources and nearly criminal extortion of labor necessary to power artificially intelligent systems, she offers a simple and keen insight captured in the sentiment of OpenAI's chief, Sam Altman:

> The most successful founders do not set out to create companies. They are on a mission to create something closer to a religion, and at some point it turns out that forming a company is the easiest way to do so.[9]

What is the core thesis of a religion based on technological wizardry and the apostles of power and wealth sitting around that altar? What is the key to their gospel? There is no question that the technical capabilities we're experiencing in this wave of AI are awe-inspiring. Individual men with personal space programs and net worths that rival the GDP of nations are equally awesome. In our quiet chambers of private unspoken envy, the wistful lament surely arises—*What if it were me?* The central gospel—unmitigated power, acquisition of fantastic wealth, and control over people's beliefs—is communion for those who possess it and a cage for those subject to it. Karen Hao highlights one of the books that is foundational to the centering philosophy among the AI illuminati about the impact of AI on society: *Superintelligence: Path, Dangers, Strategies,* by the Oxford philosopher Nick Bostrom. In an earlier defense of objectivity, Dr. Bostrom said, "I have always liked the uncompromisingly objective way of thinking and speaking. . . . Take for example the following sentence: Blacks are more stupid than whites. I like that sentence and think it is true."[10] For Dr. Bostrom to be one of the evangelical authors of the gospel of technological progress, that faith doesn't bode well for me. His is a materially different stance than Mandela's, fighting against white domination and Black domination. It harkens back to a different time when different white men astride the horse cited the Apostle Paul in his message to the Ephesians (6:5), "Servants, be obedient to them that are your masters according to the flesh, with fear and trembling, in singleness of your heart, as unto Christ." They drew on religion to power thrust in the bullwhip. It was the presentation of faith as the veil covering barbarous, dehumanizing subjugation. The biblical justification of slavery may seem an outlandish comparison to the circles of the tech sphere, with its highly polished offices and sterilized facade of objective technical merit and money. But Altman's aim "to create something closer to a religion" makes it relevant. A religion to what? Faith in what? Winning? Being first? Technological progress for the sake of power and domination? Being tall because all those around you are on their knees? The stated goal of creating a religion exposes the risks of faith in self-appointed prophetic men. The conclave of the ultra-rich and powerful technologists leading the AI charge is drawing on that theology of progress as a path to power. When power is the goal, justice is always a casualty, and domination is ever-lurking nearby. And, unfortunately, this technological march to power is coincident with the return of the more explicit racist animus embodied in the stripping of so-called DEI-related education, the declaration that Black people eat dogs, and the positioning of "Black jobs" as those of obedience and servitude—all an assault on freedom and justice.

Dangerously, the core gospel of faith in technological progress is wrapped in a false robe of neutrality, with objectivity as stitching. Those sound nice but are absent moral conviction. The apostles of technological progress tell us that their pursuits are neutral, and their work is objective. They tell us their neutrality is in alignment with the White House AI plan and that they are operating in the best interest of the country.[11] In this AI season, however, stripping so-called DEI and erasing complete education is positioned as objectivity and the bedrock of neutrality. The White

House's executive order "Preventing Woke AI in the Federal Government" helps give those faithful to Altman safe harbor for their professed neutrality. It defines neutrality in a very specific way that renders it a weapon—a cloak for a gospel of oppression and dominance.

> . . . Americans will require reliable outputs from AI, but when ideological biases or social agendas are built into AI models, they can distort the quality and accuracy of the output.
>
> One of the most pervasive and destructive of these ideologies is so-called "diversity, equity, and inclusion" (DEI). In the AI context, DEI includes the . . . incorporation of concepts like critical race theory, transgenderism, unconscious bias, intersectionality, and systemic racism; and discrimination on the basis of race or sex.[12]

Under that cloak, the core thesis of their gospel is revealed. When the examination and exorcism of systemic discrimination and racism are considered destructive, those who are discriminated against are on notice. Hence, Mandela came to see us first. He knew. The stated goal of the White House AI Action Plan is dominance. It could not be clearer. Chapter 1, verse 1: "The United States is in a race to achieve global dominance in artificial intelligence (AI)." Chapter 1, verse 2: "Whoever has the largest AI ecosystem will set global AI standards and reap broad economic and military benefits."[13] Nowhere is the idea of AI in service of a democratic, harmonious, and free society. Dominance and power are the gospel of this religion, and neutrality is its false idol.

Erasure allows for its apostles to claim neutrality as the practice of their faith. In the background of their pursuit of neutrality, they quietly submit to a form of educational and cultural cleansing. They tacitly align with—or worse, enthusiastically endorse—the idea that exposing racism and attempting to purge it from American society is so-called toxic "wokeism." In hiding behind neutrality in this way, they suggest that they are advocating for an unbiased, neutral presentation of the American story. That argument sits at the very heart of the collision between a fair and just society and the proselytizing for technological progress we're all experiencing. The makers of these AI tools suggest that they build the tools on a bedrock of neutrality. Anthropic, for example, claims to train Claude, its AI chatbot, for "political even-handedness," with a commitment to "treat opposing political viewpoints with equal depth, engagement, and quality of analysis, without bias toward or against any particular ideological position."[14] They promise that Claude will pass the "Ideological Turing Test"—describing each side's views in ways that side would recognize and support. But this framework is a sophisticated sleight of hand that weaponizes neutrality. When the Proud Boys, descendants of the American Ku Klux Klan, offer the Great Replacement Theory as justification for rabid anti-Semitism and the purging of non-whites from the country, what does it mean to not have any bias against that "particular ideological position"? When you apply "equal depth of engagement" to a debate between someone arguing for comprehensive immigration reform and someone arguing for the mass roundup and deportation of Latino communities, you have not achieved neutrality; you have legitimized state violence by treating forced removal—or, arguably, ethnic cleansing—as a rational policy position. The end of their logic lands us in a place where, from a position of neutrality, we'll debate whether unilateral accusation of terrorism is sufficient grounds for structural starvation and genocide. From a position of neutrality, we'll debate whether African American history should be taught in American public schools. From a position of neutrality, we'll debate whether slavery was good or bad for Black people. This is how neutrality creeps. Neutral becomes the clothing of complicity. It offers an anchor for a gospel of domination and sets moral conviction adrift.

These two ideals, these two distinct belief structures—one of faith in freedom and justice, the other of faith in technological progress—are standing at the gate of the second quarter of the twenty-first century. Mandela and Altman. For a child of the tattoo, it is difficult to even put their names alongside each other—one an apostle of freedom, the other of power. But they are connected in the far-reaching impact of what they represent and the profoundly different belief structures that yoke them. Strangely, they are high priests of different faiths that are not necessarily at odds with each other. The two faiths are not inherently opposing sides of an ideological war. But there is hierarchy. Technological progress can be in service of justice or in service of power. Justice, however, is an end in itself and the cornerstone of freedom. It doesn't kneel in front of technology. It serves no other master than freedom. In our current environment, where AI and technological progress are challenging our agency and belief in ourselves as people, and the American racial backlash is challenging our freedom and commitment to justice, the two are inextricably bound and evolving to become oppositional. Marc Andreessen's *The Techno-Optimist Manifesto* suggests in a section called "Becoming Technological Supermen" that "we believe that advancing technology is one of the most virtuous things that we can do."[15] With that being a foundational tenet, the current ascendency of belief in technological progress is unnerving. AI is everywhere—a kind of phantom that we can neither fully see nor fully understand—and is being used as evangelical bait by its apostles to lure us into believing in technological progress as salvation, as virtuous. At the same time, freedom and justice are under withering assault. Ironically, in some ways, they are being attacked by new AI-enabled capabilities at the hands of the very same self-proclaimed virtuous men. The descendant of the man astride the horse is oiling the bullwhip, rubbing salt in its plaits, and wrapping it in an algorithm. We've a choice now. In what do we place our faith?

When OpenAI launched ChatGPT on November 30, 2022, it changed the educational world. All of the Big Tech companies responded to the start gun of the new education arms race. The entire edusphere offered competing AI-enabled tools. The claims were big about how transformative it could be. Fundamental transformation of the practice of teaching. Teachers would be able to use AI-enabled tools to teach in radically more efficient and creative ways. They would have the benefit of the synthesis of the entire internet to help them configure, cohere, and deploy content in new pedagogical ways. Companies promised that these tools would dramatically tip the balance back to teaching from administrative tasks for teachers, allowing them to be teachers again. The claims about learning were equally grand. Tech companies offered that ChatGPT and AI-enabled tools like it would be revolutionary to individual student learning. The bold claims included that all students could have personalized technical tutors who would be responsive to their individual learning needs. The combination of radical improvement in teaching and learning, with a responsible approach, would finally help speed the closure of persistent American education gaps. In an article titled "The Future of Learning: How AI is Revolutionizing Education 4.0," the World Economic

Forum offered, "Imagine a future where all students receive personalized support and inclusive learning opportunities, build stronger connections with teachers for enhanced guidance and receive apt recognition and evaluation of their achievements."[16] It was all grand.

Despite the grandeur, in education, the different faith traditions—Altman and Mandela—lead to very different ends. Faith in technological progress leads to what we currently see. It is a kind of turbocharged techno-optimism speeding us toward a professed educational utopia. We are besieged by promises of enhanced teaching, learning, and enlightenment through AI-enabled educational tools and eye-popping announcements of billion-dollar investments in education: "Microsoft Pledges $4 Billion Toward A.I. Education," "Google Spends $25M on AI Training for Teachers."[17] At the core of these claims is "unlocking the potential of AI in education." One of the more highly regarded summits on AI in education, hosted by the Stanford Institute for Human-Centered Artificial Intelligence (HAI), promoted at their 2025 summit a conversation to "unlock the potential of AI to personalize learning, empower teachers, and bridge educational divides."[18] The Bill and Melinda Gates Foundation offers hope of "Unleashing AI's Potential for Equitable Academic Outcomes and Economic Opportunity."[19] Edtech companies brokering for supremacy in AI in education are making lofty claims about the millions of teachers and students who are beneficiaries of their various approaches. One company, aptly named AI for Education is "Dedicated to AI Literacy for 1 Million Educators & Students."[20] Their approaches are all variants of a similar theme. They offer some version of facilitation of AI literacy for teachers and students that will transform the relationship between teaching and learning.

AI literacy is the ability of teachers and students to use AI-enabled tools, to be able to effectively navigate the prompts and assess the veracity of the results offered. According to Microsoft's *2025 AI in Education* report, "AI is evolving from being a simple time saver and productivity booster to a 24/7 assistant and force multiplier. It can ready students to tap into their creativity to bring projects to life."[21] It also means teaching both teachers and students the applications of AI in the world. The facilitation and professional development offered will help teachers reclaim their time from the onerous tasks of lesson planning and scope-and-sequence visioning. They offer creative ways to incorporate AI tools into pedagogical practices, to lean into the technology and unleash its potential to improve their practice of teaching. According to aiEDU, one of the more popular organizations in the AI edusphere, "our trainings are designed to help build AI Readiness and empower students with AI literacy."[22] Many of the companies and organizations also offer evaluative tools for teachers to develop savvy in determining whether and how their students are using AI tools. Behind those claims, they often offer motivational dystopian prophecy that the world will soon be divided between those who use AI tools to improve whatever their craft is and those who do not. With a tone of foreboding, they offer that education is not exempt from that divide; therefore, it is incumbent on them as teachers, or district and state leaders, to engage in this effort, lest they be left behind. Parents will be demanding it, asking what the district is doing to prepare their children for an AI world, and employers will be asking how they are educating their future workforce to be compatible with the AI demands of the future. Faith in technological progress as a catalyst to improving education has clearly taken hold and is spreading with evangelical zeal.

So, what does this really mean? Where does this Altman faith in technological progress lead? Where are the technological supermen leading us? What are these grand claims of universal improvement really saying these strategies and tools are capable of doing? Making claims about the improvement of millions of teachers and students, as many of these edtech companies do, may be true, but for those of Mandela's faith, they don't inspire. In a national public education system of more than three million teachers and nearly fifty million students, where Black students have had the breath of their ambition and talent choked out of them for generations, these claims are notably devoid of justice. Code.org, one of the most popular computing education advocacy organizations and a darling of big tech philanthropy, boasts more than 102 million students globally touched by their Hour of Code and other programs, and more than three million teachers using their platform.[23] They offer that their Hour of Code sparked a generation, and Hour of AI will define what's next. The Khan Academy, arguably the gold standard for online teaching tools, suggests that more than six million teachers globally have turned to it. They now offer Khanmigo, "your free, AI-powered teaching assistant," and they are positioned to be among the leading thinkers about the future of learning.[24] I suppose there is sincerity in their claims. Replicability and scalability are the kinds of words that, in the world of philanthropy and corporate giving, mean something. They may be thinking that bringing learning tools and opportunities directly to individual students and teachers is its own form of justice—a direct tactic to circumvent the constrained and abusive school systems themselves. That is a valid approach. If you think you can't change the system, develop a work-around. Revolutionary subversion has its place, for sure.

But the reality is what it is. School is school, and Black students are still Black. Black students are still both dependent on public school and abused by it. Math is Power has come and gone. The math caste fought back and is still standing undisturbed and undefeated. Code is Power has come and gone. The CS caste fought back and is still standing undisturbed and undefeated. The literacy fight is never-ending, yet caste seems to have infinite endurance. The National Assessment of Educational Progress (NAEP) results in 2024 showed that 56 percent of Asian eighth graders were proficient in reading. White, Latino, and Black students fell in line according to their educational rank—38 percent, 21 percent, and 14 percent, respectively. Despite the more than 100 million students sparked by Hour of Code, and the millions of teachers empowered by the Khan Academy, only 3 percent of students at the most elite computing institutions in America are Black.[25] Sal Khan, the founder of Khan Academy, went to MIT and studied electrical engineering and computer science. In the first admitted class after the *Students for Fair Admissions v. Harvard* Supreme Court decision in 2023 ending affirmative action in higher education, MIT's class of 2028 dropped to 5 percent Black students, down from 13 percent.[26] Hadi Partovi, the founder of Code.org, has a master's degree in computer science from Harvard. Harvard's master's program in computer science is comprised of less than 2 percent Black students. Sam Altman, the current papal leader of this faith in technological progress and optimism in education, went to Stanford. There, too, the percentage of Black students in the class of 2028 dropped to 5 percent from nearly 10 percent in the class of 2027.[27] These men, and their faith in technological progress, are proselytizing from a conclave Black people can't access—a conclave increasingly donning the symbolic sign: "No Blacks or Foreigners Welcome." They're offering a savior's gospel that provides no path to the altar from which they're delivering their sermons. For now, at least, in the emerging AI world, power resides in the elite corners of the tech enterprise. That is the hard truth. Justice is fighting for access to that—the epicenter of power. Laying a tech-enabled blanket over a profoundly segregated and unjust school system isn't that fight. It merely hides the brokenness and places hope in shiny things. It doesn't carve a path to justice. When technological advancement is seen as virtuous, justice can be a hindrance. In the best interpretation, that approach suggests that segregation and injustice are constant variables of a larger system to be designed around—hoping that tech tools can breathe life into a sea of choking Black students. At worst, that approach is an opiate designed to distract and lure us away from the actual fight with the promise that AI will make the crooked places straight.

The American assumption, whether spoken or not, reflects American caste, and as Derrick Bell pointed out, Black people are the faces at the bottom of the well. If the task is to merely help everyone do what they're doing a little bit better, the assumptions remain intact and caste remains unchallenged. Without addressing the assumption, AI is Power renders itself verisimilar to Math is Power and Code is Power. Just because AI is an extraordinary technological leap forward doesn't mean that it will do anything to break the rank order of America's racial groups in education if that goal is not explicit. It will not happen. Technology is not capable of that on its own. And claims of not wanting to have so-called "at-risk" and "marginalized" students fall farther behind are nice sounding, but the assumption remains that they will stay behind, just not as far. What the defeated Math and Code are Power campaigns had in common was that they did not confront the assumption. They attempted to assert power for a group within a system that renders them powerless. In both cases, the focus was mostly on the classroom—the teachers and the students. As teachers like to say, "Within these four walls, I am in control." Give the teachers better tools, better pedagogical approaches and curricular content, and give the students better learning devices and better incentives, and educational progress is born. While that logic is reasonable, the injustice within which the logic sits is staggering. How will an incrementally better approach in the hands of a struggling teacher with struggling students in a struggling school in a struggling neighborhood put a dent in caste's armor? Here, too, the conclusion is not to be fatalistic but realistic. The grand claims of what AI is going to do for education rely too heavily on a belief in false prophets that see technology itself as a virtue, not justice. To assume that the least enabled students and schools will somehow move faster and catch up to the most enabled students and schools because of AI tools is naïve, if not silly. Being enabled is the substrate that gives anchor to educational innovation. Those students and schools that are already high functioning are the very schools that will be most capable of experimentation and most receptive to the innovation that invites experimentation in the first place.

The Constellations Center for Education in Computing that I run at Georgia Tech has taught me this lesson clearly. One of our early goals was to build high-quality Advanced Placement computing courses for schools within Atlanta Public Schools. The entire thesis is to offer the nearly 75 percent of Black students in APS, the native children of Atlanta, a shot at the tech-based economy that the city is building. In our narrow lane, we're trying to help forestall the massive out-migration of existent Black communities that has befallen every other major American city that has sown big tech growth and its associated gentrification in the last twenty-five years.[28] As I've mentioned, the students in the all-Black schools have woefully underdeveloped math skills. Their school administrators, therefore, understandably operate in triage mode. They see the need for what we do and want computing education for their students, but they must address their much more fundamental needs first—basic mathematics and literacy. The superintendent of APS has even issued them the charge: *Back to basics*.[29] But what even are the basics in schools where collective high achievement was so long ago that no one remembers when not doing well was a shame? They are structurally constrained, which limits our ability to be present in some schools, let alone be innovative. By contrast, in the one school that is majority white, they have so many students taking AP courses that they have a built-in study hall period for a single credit just for those students to keep pace with their work. They're innovative even in their accommodation of high achievement.

It is relevant to offer how I learned about this class—the kind of detail that is lost in the popular discussions about AI literacy and its potentially turbocharged educational effects. Not only are the schools in Atlanta Public Schools nearly completely segregated, in the three traditional high schools where white students attend, there is a low-grade segregation among the teachers too. Black teachers tend to only teach Black students. Where there are high concentrations of white students in advanced courses—IB and AP—they tend to have white teachers. In the one school that is nearly half Black and half white, I was in the hall and noticed a large class of white students with two Black students, and the teacher was Black—an anomaly in the system. Looking in the window, it also seemed that the teacher wasn't doing anything. I knocked to ask. The teacher came to the door and stepped outside the room briefly. "Hey good Brother," I asked, "what class is this?" He said it was AP Study Hall. He explained that it was an in-school study hall for students with three or more AP classes to be able to do some of their homework during the day. Their parents were complaining they had too much homework and it was eating into their family time, so the school accommodated them and let them do their homework at school. The kids were doing work from their various AP classes; that's why he wasn't doing anything. He was just a chaperone. I asked why there were only two Black students in the class. Without pause he said, because there are only two Black students in the whole school taking three or more AP classes. "Brother," he said, "they got me in here babysitting Midtown white folks' kids. You know how it is." I do.

In that very school, our ability to build the structure for AP computing is met with enthusiasm and creativity. The administration suggests feeder courses, integration with parallel courses in technology where they think computing could enhance students' appreciation of the applications of computing. As we're moving into the AI world, the enthusiasm has only intensified. They come to us asking for more help, more integration with what Georgia Tech faculty are doing, to expose their students to the cutting edge. Our challenge, of course, is that the Black students in the school are hidden behind a wall of underachievement and the ambiguous barriers to being recommended for AP courses, or they are steered to the equivalent of vocational education, where the white students are glaringly absent. This is the street-level reality that doesn't appear in the high-minded discourse of techno-optimism and its unleashed potential in American education. This is the daily grind of race that the disciples of technological progress fail to acknowledge at best, or at worst, seek to reinforce.

Atlanta Public Schools offers a non–Advanced Placement version of the AP Computer Science Principles course. In 2024, across APS, 94 percent of the students who took that course were Black, and only 1 percent were white.[30] By contrast, the gold standard of AP computer science is the AP CS A course, where students learn Java. Across the system in that class, 50 percent of those students were white, and 34 percent were Black. That is in a school district where nearly three-quarters of all the students are Black and less than 15 percent are white. In the introductory course of the vocational education track to computing, Introduction to Digital Technology (IDT), nearly 85 percent of the students were Black and less than 10 percent were white. And the general anecdotal understanding among those familiar with the vocational Career, Technical, and Agricultural Education (CTAE) pathways in the Atlanta Public School system is that the IDT course is virtually useless. That is where the fight is. That is the justice fight. Those are the tell-tale patterns that indicate injustice is lurking ever near—a quiet, systematic, perennial pulling of the Black kids to the bottom of the well. The advanced capacity of the school and the progressive approach to educating the white students will surely accelerate their AI literacy and readiness—the potential of AI in education will indeed be unlocked, and we're watching it happen. The questions about the relevance of learning a specific computing language in the face of generative AI tools are already underway. What does computing literacy look like in the age of generative AI? Questions like these are only possible to answer when basic literacy is an assumption, and at least here, they're being asked de facto with white students in mind. Technological progress will help them for sure. But it is educational injustice that keeps Black students in Black schools locked out—and Black students in integrated schools as witness to progress but not beneficiaries of it. Under those circumstances, the idea that AI literacy will help Black students "catch up" with their white schoolmates is meaningless communion to a false virtue.

Here, perhaps, it is relevant to restate that I was born in 1972, and I am of the first generation of fully free Black people in American history. Prior to the passage of the Civil Rights Act of 1964, it was legal in America for states to exercise their rights to explicitly discriminate against Black people. When Nelson Mandela emerged from prison and arrived in Brooklyn in 1990, I was eighteen years old. When I left home to go to college, it was still legal in South Africa to stop Black people at gunpoint and demand they show a pass for permission to be on the street. That perspective helps anchor the significance of justice and offer a broader view of the challenge at hand. The assumption that the mere deployment and penetration of tech-enabled tools will guide us to a more just and equitable educational landscape is disconnected from history. There is no resistance to technological advancement, just a periodic call for regulation. We assume that it is beneficial, because it is. There are nefarious uses and real dangers for sure, but in the main, over the long arc of human history, scientific and technological advancement have helped make living easier. I am a former program officer of the National Science Foundation of the United States; I believe that premise deeply. Technology is, in fact, a cornerstone of human progress. That is, in part, why it is so easy to pretend technological progress is innately virtuous. But the ability to inspire awe through extraordinary technical accomplishment and wealth can be misconstrued. Being technological supermen doesn't make them, the nature of their claims, or their work virtuous; it just makes them incredible.

History is laden with incredible men who have left us incredulous and deeply wounded. In this modern era, the era of the end of the second Reconstruction, the requirement of a call to justice and the conjuring of real virtues couldn't be clearer. That clarity holds when the tech onslaught is bearing down on education. Assuming that technological progress will benefit education is easy, whether or not it actually does. Establishing the justice that will make that technological progress fair and just is not. Given the gravity of the threats we face, the explicit aim of subjugation and erasure, and a return to a time gone by, where the question of whether Black lives matter was settled on the long branch of the lynching tree, justice is what matters most.

Laying out the gravity of that point of view, the view of the Mandela philosophy, is missed in the glee and enthusiasm about the convergence of AI-enabled tools and education. It is often seen as hysterical and heretical in tech education circles today, if seen at all. They scorn the mere mention of the words—justice, equity, fairness, history, and freedom. In today's America, this view is a kind of infidel bound to the stake to be stoned with DEI rocks. It doesn't align with the excitement of unlocking the potential of AI. Neither does it match with the defanged language that fits through corporate legal and public relations filters to highlight "for all" effects that will offend no one and mean not much to anyone. It is the view from outside the conclave. It is the view of Langston Hughes's darker brother, from the kitchen, when company comes. It is the view from the Black teacher babysitting high-achieving white students, knowing that Black students are trapped in the basement gasping for air. It is a view that is inconvenient to the tech advancement orthodoxy. This point of view conjures memories of lessons that the tech supermen perhaps never learned, likely weren't affected by, and therefore don't remember—memories of a timeline that outline the actual fights for educational justice. Memories of fights where evidence of impact was assassination—not the next hundred million users. Those are the grand and historic battles for fair and just American education.

The justice fight is a fight because there is resistance—there is opposition to the stated goal. Not only is there opposition; there is also a guilty party. Injustice, like marginalization, is not a passive occurrence. Someone has to do it; someone is guilty of having done it. People or systems must be held to account for their transgressions. That is a core difference between the two ideologies. In the excited thesis of technological progress as a catalyst for fair and equitable education, there is no assumption of resistance. The apostles of this faith assume goodness in themselves and in the educational system that created and benefited them. They've never been asked by the descendant of the tattoo: How do you plead for rendering the scar on my back—guilty, or not guilty? They've never had to contend with being spit on for trying to go to school or looking on from the basement as the three-course AP students upstairs leave them in the last century. They are the three-course AP students. The Mandela thesis, on the contrary, understands that justice requires a fight. Assuming goodness in a system that has repeatedly forsaken you is folly. Asking us to trust a system upon which there is no evidence for us to base that trust is dangerous. We trust our memory, and it informs us that educational justice requires a fight.

The *Brown v. Board of Education* effort to desegregate schools was that kind of fight. The extraordinary resistance from white communities to having Black children attend school with their children is evidence of that. It was an attempt to right a historical wrong, segregation, and achieve a fair and just educational system. Large segments of white society chose to fight against that. The *Regents of the University of California v. Bakke* Supreme Court case, which introduced affirmative action in higher education, was that kind of fight. Race had been the single central factor in denying Black people access to higher education for so long, it had to be equally considered in righting the original wrong. Here, too, large segments of white society felt that, despite having used race as a cudgel to deny Black people access for so long, we are now to believe that we magically live in a race-blind society—we are now meant to trust in their self-proclaimed salvation. Again, dangerous. The paths to and from both of those epic American education justice battles are littered with bodies and acrimony. The 1964 Freedom Day boycott of New York City public schools, when nearly 470,000 students, almost half of all the students in the system, walked out to protest the horrible conditions in Black schools, was that kind of fight.[31] The students themselves saw from the basement that what was happening to them was criminally unfair, and they fought against it. It was a fight based on their actual experience, not the proclamation that their conditions would improve by miraculous technological decree.

These were the kinds of fights that centered justice spoken out loud—the righting of American wrongs—and confronted the darkness of the nation's past. These were fights anchored in Mandela's thesis of a free and just society. They offended, disturbed, and provoked because the path to justice in an unjust society must. When justice is the goal, a fight is guaranteed, because injustice fights back; it doesn't concede. Pretending that technological progress is virtuous and can serve as the central tenet of a religion is offensive to children of the cross and the lynching tree.[32] Our wounds from the fight for justice run so deep that our memory is its own sanctuary—and we recognize false idolatry when we see it.

CHAPTER 6

EDUCATION VERSUS TRAINING

The dominance of computing and all of its power and awe and potential and threat in this century has reshaped our collective view of education. STEM education—science, technology, engineering, mathematics, and computing—has battled its way to the top of the mount. In some circles, STEM education is the only education. The universal applicability of STEM-related skills, especially computing, has become the foothold of its ascension. In order for students in American schools to pursue any interest they may have, from arts to law, from medicine to music, there is a need for technical skills, if not computational fluency. At least, that is how the argument goes. As such, the emergence of STEM schools, of STEM certifications, of STEM funding networks, affiliates, camps, scholarships, and advocacy groups has changed the modern landscape of education. Much of the culture shift has been driven by the ubiquitous presence of computing. The arrival of AI has only turbocharged the inertia of that already powerful movement. It is a simple and reasonable logic that has welcomed the business case language into education. What is the return on investment? What fields lead to the highest-paying jobs? What is the average starting salary per disciplinary category—English versus computer science? There are obvious practical reasons to care about such things in an advanced society such as ours, which has a woefully poor social safety net. Under that broad brush, the objective of education has shifted from being enlightened to being enriched—the pursuit of passion smothered by the need to survive. This need is universal, of course, so getting the skills necessary to secure a high-paying job has increasingly challenged the broader mission of education. It forces us to question whether education has become a complex job-training program or a vehicle for an enlightened citizenry and whether the two are mutually exclusive. At the center of these converging pressures is the difference between education and training.

The collision of AI and anti-DEI forces on education makes the distinction between education and training an important one to consider. AI is putting enormous pressure on education to incorporate AI literacy in preparation for the future of work. It is tipping the scale and amplifying the business case argument that we need an AI-literate workforce and that the education system needs to provide that. That is a practical motivation. It is reasonable for the US education enterprise to be responsive to that goal in service of the national interest. The risk is that the emphasis is so overwhelming that it strips education of education and transforms it into training. It is also a convenient shift in the season of anti-DEI. In this period of backlash, those districts that buy into the anti-DEI rhetoric can emphasize the utility of their stripped-down version of education as being indoctrination-free, skills-only education. Other districts that are forced to bend can arrive at the same conclusion, albeit behind the justification of compliance. It is alternatively insidious or safe, depending on your point of view. But in either case, it forces consideration of the distinction between education and training.

On its face, the difference between the two can appear trivial. But when you're building entire educational systems around them, the distinction matters. In articulating the difference, the words can get very fancy very fast—epistemological rationale, erudition and enlightenment, citizenry and democracy versus vocational readiness, skill-based practicality, relevant skills, and economic mobility. The intensity of the excitement about AI in education not only runs the risk of luring our attention away from educational justice, it increases the likelihood of greater emphasis on training. That is one version of what gutted education looks like. Notwithstanding the academic debate that educational theorists might have, there is a simple distinction that is meaningful in the context of AI in education.

Education is the development of intellectual rigor and critical thinking. It offers continuous inquiry, curiosity, and the ability and drive to learn and acquire knowledge. It provides for a yearning to understand why. Training is oriented toward the development of particular skills that are applicable to specific tasks. It has a more pragmatic goal of occupational readiness. It leans toward certification and away from erudition. It centers on how things work rather than why they do. They are both important in context, but they are not the same thing. Education is more comprehensive than training. If school systems are oriented around training, students graduating from those systems will be well trained in certain skills but not necessarily appreciate or question the reason the skills they've been taught are the skills they've been taught. Who made that decision and why? They're guided to know how but not why. Education, on the other hand, doesn't preclude the acquisition of skills. It incorporates the practical acquisition of skills in the broader goal of acquiring critical thinking and the ability to learn how to learn. In that sense, it provides for the critical anticipatory foresight to know when an acquired skill may become obsolete and the ability to learn new ones. It also provides for broader cultural fluency in literature and history that situates learning in historical context.

These differences are not pedantic when applied to massive systems like public education in the United States serving nearly fifty million students, or the nearly twenty million students in postsecondary education. What happens when training without critical thinking is amplified to tens of millions of students? The coincident executive orders—one supporting AI in education and the other blocking DEI in education—is an attempt to create this very scenario. Undoubtedly, the advocates of the anti-DEI ideology would argue that education has nothing to do with DEI. They would argue, on the contrary, that education is distorted and defamed by DEI. At this point in the American story, however, it is obvious that there is no particular definition of what DEI is. It is used as a kind of all-purpose indictment to justify any manner of retribution and attack. It is not dissimilar to declaring someone a terrorist as the only justification necessary for destroying them. However, there is a special denigration of Black people that is core to the anti-DEI effort. Nikole Hannah-Jones's *The 1619 Project* is an example.[1] Her work is a historical account of the foundational role slavery had in the construction of the United States. Despite the inarguable factual basis of her work, and the direct lineage of her own family involved in the narrative, it was branded as DEI and is a banned book and banned from curricula in several school districts.[2] Having weaponized DEI, it gives license to the adherents of its populist, if not racist, ideology to just declare, indict, and erase.

Historical context is fundamental to a broader education. Within it sit approaches to historical eras and the culture, language, literature, conflicts, and circumstances that form our current national fabric. That kind of information helps position our understanding of our station not only in the contemporary American story but in contrast to other eras and in the world. It helps us see social hierarchy and the historical rules and circumstances that create social patterns. Understanding the cultural anchors and traditions, for example, that render Asian students so far ahead of white students in academic performance in a society set up to advantage white students comes from examining history and large social clues. Waxing poetic about the importance of a liberal education runs the risk of sounding wistful, as if it is already a thing of the past. There is no question that it is under attack by the anti-DEI advocates, but even despite them, the wavering between education and training is happening. The dominance of the business case and the need for efficiency in developing a workforce prioritizes training. Education is increasingly viewed as inefficient against the demands of a trained workforce that needs to be ready yesterday. Education itself is taking its place on the long list of rich societal features sacrificed on the altar of efficiency. The surge of crudeness and the bludgeoning of intellectualism in America is, sadly, another byproduct of an overemphasis on training and the undereducation that goes with it. But it is important to point out the difference between education oriented around that larger intellectual goal of having an informed citizenry and the training of students to be AI literate, certified, and prepared for the workforce.

AI in education does not necessarily equate to education as training. The counterargument would be that AI-enabled tools are being used to enhance all subjects of education to enrich the entire experience—that AI-enabled tools can be used in pursuit of the broader goals of education. There is a component of the effort that is focused on that. Yes, there is and must be. Questions and experiments about how to improve pedagogy and content across all disciplines are underway in many schools and districts. While that may be the case, the broader effect of AI in education is more intense preparation of an AI-skilled workforce—however loosely defined that may be. The president's executive order is clear in its objective, "Early learning and exposure to AI concepts not only demystifies this powerful technology but also sparks curiosity and creativity, preparing students to become active and responsible participants in the workforce of the future."

That is an explicit orientation toward work. The goal is to expose students to technology and train them to be workers. The pragmatism of that goal is not lost. But having a population of workers that may not be armed with the other components of a complete education renders a population that is easily mailable. In the United States, that means a population that can easily slip back into the default orientation of the country, where racism is foundational. That is why the difference between education and training is important, and how AI in education is connected to this bigger national challenge.

The tendency in STEM education is to pretend that it is wrapped in objectivity because we associate that quality with mathematics and science. But the broader issue is the use of STEM education and its cloak of objectivity to crowd out the importance of a full and honest education. That use is both politically and culturally charged. It is dripping with subjective, subversive goals. The wholesale emphasis on AI in education uses the pragmatism of training as a draw to devalue comprehensive education and minimize the size and voice of an active, critical citizenry. AI certification surely doesn't include learning about the Trail of Tears, for example. Hence, a well-trained AI-certified worker may not see the parallel with the current attempt to round up eleven million people living in the United States and forcibly move them somewhere else. They may work in a company seeking to employ powerful AI-enabled tools to use biomarkers to identify so-called "illegals" to seize them from their homes, schools, and churches and send them to detention camps.[3] They may do so not knowing that when history looks back to judge such things—the rounding up of Native Americans, the rounding up of First Nation people, the rounding up of Japanese people, the rounding up of Chinese people, the rounding up of Black people, the rounding up of Jews—the verdict is always the same: guilty. The contemporary justification, whatever it is, never stands the test of time. They may do so, not realizing the historical pattern, because they've been trained on how to execute the task but not to ask why do it in the first place.

The attack on DEI in education is an attempt to remove access to the fullness of the American story, which is core to a complete education and the motivation of the question: Why? At best, the goal is to simply replace education with training and leave students to fend for themselves to learn about the world in the algorithmically curated way that will now occur. At worst, it is an intentional effort to replace actual history with a revised version that reduces the significance of Black people and others over the course of the American arc to footnotes. In some ways, that is where the concept of marginalization comes from. To be marginalized is to have your story pushed to the margins as mere edits to the actual story. It is an active process, marginalization. Someone has to marginalize you. The attack on DEI in education is the process of marginalization and the subsequent subjugation of people. The emphasis on training under AI in education offers cloak and cover for that process. Under these circumstances, the distinction between education and training is far more significant than mere definitional differences.

The first Trump White House issued an executive order to establish the 1776 Commission in response to the surging inertia of the 1619 Project.[4] Nikole Hannah-Jones's *The 1619 Project* was so comprehensive and compelling about the actual American narrative, it was being translated into curricula and beginning to be adopted by school districts around the country. So, it was a showdown: 1776 versus 1619—American history told from the bullwhip versus American history told from the tattoo. The central argument of the 1776 EO was the following:

> In recent years, a series of polemics grounded in poor scholarship has vilified our Founders and our founding. Despite the virtues and accomplishments of this Nation, many students are now taught in school to hate their own country, and to believe that the men and women who built it were not heroes, but rather villains. This radicalized view of American history lacks perspective, obscures virtues, twists motives, ignores or distorts facts, and magnifies flaws, resulting in the truth being concealed and history disfigured. Failing to identify, challenge, and correct this distorted perspective could fray and ultimately erase the bonds that knit our country and culture together.

In the 1776 interpretation, who were the men and women who built the country? Was the master to the slave a villain or hero? Could the slave to the master be either, given his status as thing? Was Nat Turner, for example, hero or villain? Why did he even need to be a hero, given the country was already founded and the virtuous founding documents already written? Were the white men who hung him and skinned him heroes or villains? The assertion of the 1776 order is that the virtues of this nation ought to lead us to conclude that Nat Turner was a villain and the men who beheaded him were heroes. In the Black world, in the world from the point of view of the 1619 Project, Nat Turner is a revolutionary legend. Against near-certain death, he fought for freedom, and in 1831, he led the largest revolt of the American slave era. He is more than a hero; he is a liberatory deity. In some scholarship, he is considered one of the most consequential Black people in American history.[5] The significance of the different points of view is that the adjudication, the disentanglement of the American story can only come through a complete education. Because who we are as a nation matters; it is essential for us to understand both how we have become the nation we are and why. The whole story of Nat Turner in this environment can be sprinkled with anti-DEI dust, proclaimed as the radicalized idolatry of villains, and summarily banned from public education. The ability to grapple with this very fundamental understanding of the nation's history and its instincts and origins comes from education. It is more than practical; it is existential.

It is a national crisis that, so far into the American story, approaching the 250th anniversary of the country, when the shape-shifting power of AI is descending on us, this case of basic history remains unsettled. The smothering of education as a precursor for subjugation ought to be in the nation's past. But yet, here we are, just as we were at the end of the first Reconstruction. Caste is fighting back and reasserting itself in the aggressive diminishment of Black people. In the preface to *The 1619 Project*, Hannah-Jones lays out the challenge:

> Black people, however, were largely absent from the histories I read. The vision of the past I absorbed from school textbooks, television, and the local history museum depicted a world, perhaps a wishful one, where Black people did not really exist. This history rendered Black Americans, Black people on all the earth, inconsequential at best, invisible at worst. We appeared only where unavoidable: slavery was mentioned briefly in the chapter on this nation's most deadly war, and then Black people disappeared again for a full century, until magically reappearing as Martin Luther King, Jr., gave a speech about a dream. This quantum leap

> served to wrap the Black experience up in a few paragraphs and a tidy bow, never really explaining why, one hundred years after the abolition of slavery, King had to lead the March on Washington in the first place.
>
> We were not actors but acted upon. We were not contributors, just recipients. White people enslaved us, and white people freed us.[6]

Her theme, that in this version of American history "we [are] not actors but acted upon," is consistent with the caste theme—creator versus consumer, part of the main text versus living in the margin, a school of failing algebra versus a school of AP Calculus, AI in education versus education for AI. To magically disappear is to be the faces at the bottom of the well, the children of Sisyphus, the children trapped in the pattern. The only possible path to challenging that order begins with seeing it. Therein lies one of the core functions of education that is materially different than training. The broader view of the social position in which our functions sit comes from education. The ability to see the correlation between social function and position is born not just of experience and second sight and an instinctive yearning for something more but of intentional education. When the US senator from South Carolina, James Hammond, suggested in 1858 that the South had found a people aptly suited for menial labor and called them slaves, he was outlining a particular social structure and assigning Black people to the bottom of it. He was assigning Black people then to what today the president refers to as "Black jobs." The simultaneous ban on Black literacy and the absolute illegality of education in the slave era were to block Black people from learning the broader and more global philosophy of freedom and enlightenment. It was intended to ensure that they did not conspire with each other or learn the thinking that their white contemporaries used as justification for the Revolutionary War for gaining their own freedom. The nineteenth-century banning of Black literacy and education has a haunting similarity to the education that yields a 14 percent literacy rate and a 9 percent math proficiency rate among Black eighth graders today.[7] While Black education is not illegal today, the outcomes for Black students approximate what they might be if it were illegal. The outcomes are as if students had to learn through subterfuge and candlelight. In the James Hammond days, the banning of education was to ensure that Black people wouldn't see the very contradiction written into the founding documents that is so central to the

1776 executive order—that the virtuous Founding Fathers asserted that all men are created equal, yet counted a Black person as only three-fifths of a person. The result of perennial undereducation and overemphasis on training today may be the same: to keep Black students in James Hammond's allocated slot at the bottom, as the mudsill of society.

However, the banning of education did not stop the ongoing process of training the enslaved in the most efficient techniques of blacksmithing, cotton harvesting, carpentry, masonry, and glass blowing. Hear the origin theme of Governor Ron DeSantis's state educational standard[8] that claims slavery offered Black people training on beneficial skills, as well as his coincident declaration that "Florida is where woke goes to die."[9] The white master class was comfortable offering training to Black people—trained and asleep in a perpetual state of illiterate lethargy—but not education. That is the flag, a glaring indicator of the difference between education and training. If education will lead to the slave's more aggressive agitation for freedom, but training will not, the two are not equivalent. One is a threat to an oppressive system. The other is not. The allowance of one and not the other, among other things, provided for a permanence of caste to keep slave as slave and master as master. While that may appear to be a crude comparison of language and thinking from a bygone era, the parallel is clear. The attempt to strip DEI from schools—or in Floridian language, to kill wokeism—is in the spirit of banning education from the slave house. The historical origin is hardly obscured. To the extent stripping DEI from schools is a reincarnation of the sentiment of the man on the horse, the similarity is clear. How dare Black people assert? Be trained and be quiet is the strategy—leave education and the grand thinking to us. If left to the legacy of the man on the horse, our contribution to the American story will be banished to villainy and condemned as distorted history. Banning *The 1619 Project* under the banner of DEI is symbolic evidence of that very sentiment—to keep well-trained but uneducated slaves. It is precisely the same thing.

More than that, banning DEI in education is an erasure. The arguments of indoctrination and parents' choice and white Christian values, of lurid LGBTQ accusations, of America's perfection and its firstness—they will abound. I am neither persuaded nor moved by any of that. They are complements to the theme of putting Black people back into our place of subordination. The banning of DEI in education is an extension of the destruction of the Black Lives Matter mural in Washington, DC. Not only is there a symbolic destruction of the idea that our lives matter; the education that includes the fullness of the Black story in and of America is being destroyed too. The dismantling of Black studies programs in universities, the undoing and under telling of the significance of slavery and racism and their long tentacles into the current affairs of the national condition all lead to the same end: erasure.[10] The coincidence of promoting AI in education for the workforce while forcing universities to jettison their programs that offer critical race studies and threatening the entire K–12 public education system to exorcise any education about America's fraught racial foundation is an attempt to strike the story from the national memory. If students cannot learn about it, in time, it fails to exist as memory morphs to myth. Or the story returns as some reframed romantic blip burnished of barbarism. In that telling, the story is of American heroism and the verdict—Nat Turner was the villain and the men who killed him were the heroes. This is the process by which skills supplant knowledge. These are the extraordinary stakes in the tug of war between training and education.

I studied mechanical engineering as an undergraduate and then experimental solid mechanics at the master's level at UC Berkeley. Afterward I earned a PhD in science and technology policy from Georgia Tech. The title of my dissertation was "The Duality of Innovation: Liberation and Economic Competitiveness." My family is all from Guyana, and I am the first to be born in the United States. My master's thesis at UC Berkeley was in a very specific lane of solid mechanics called tribology, the study of friction and stiction forces. When you get out of the details, it is easy to see that in nearly every aspect of our lives there are surfaces constantly rubbing against each other—the tire on the road, the brake caliper on the rotor, the bike chain on the cassette, bodies during sex, the left and right inner thighs of the marathoner, the spinning gears of the smoothie blender, the articulating components of an artificial knee. Needless to say, it is important to understand how friction at the interface between surfaces in relative motion works for all manner of reasons. My master's thesis had to do with a now-obsolete technology: the interface of the read–write head landing on the surface of the disk in a computer's hard drive. In those days, there was an arm within the hard drive of computers that positioned a very small electromagnet above the disk that contained the magnetic patterns that stored information. When the disk spun, the read–write head "flew" above the surface of the disk at a height of approximately three to five nanometers on a cushion of air that rotates with the disk. For comparison, a human hair is approximately eighty thousand nanometers thick. When the computer is turned off, and the disk stops spinning, the head moves to a landing zone to land. That is where my work was. What happens when the head touches down on the disk to land? Much like when an airplane lands, what is the material of the tire and of the runway, and how do the two interact when the still tire hits the asphalt at nearly 200 mph? And what happens when you turn the computer on, and the disk starts to spin again? How does the head overcome the stiction forces and take off without being ripped off? In order to test such things on the nanoscale, we had to conduct experiments in the night when the building vibrations were at a minimum. It all felt very smart. High-end technical training at one of the premier colleges of engineering in the world.

During that time, I went to visit my great-aunt in Guyana, Aunt Lynette Dolphin. I spent a lot of time with her growing up, and she took a particular interest in me—the first experimental child of the Guyanese migration to North America in my family. I was a hybrid—neither fully American nor fully Guyanese. I am "Guyanese extract," as the saying goes. My parents sent me "home" often to stay connected. Aunt Lynette was an important feature of home and an important figure in the emergence of independent Guyana. She was exceedingly proper. The kind of woman who, in her presence, you stand up straight, speak properly, and all your manners are on high shine. She only wore African dresses and solid gold jewelry. She was in London studying at the Royal Academy of Music during the Blitz in 1940; the stress turned a streak of her hair gray, and she dyed her hair black ever since. She was an absolute mix of Africa and colonial Guyana. The country gained independence from the British in 1966. Her influence over the nation's transition was rooted in culture and bringing the folk traditions of Guyana and of the Caribbean region more broadly to bear on the future of the country. Her core theme was that it is our cultural ties that keep us connected to our African roots during this long sordid story through slavery and colonization. Various Europeans that upended and subjugated our lives—the Dutch, the French, the

British, and now, the Americans, but cultural ties persist. Her message was that protection of our culture is protection of our identity. Among several other things, she was chair of the Department of Arts and Culture, which became the National History and Arts Council upon independence. For all of her work, she was awarded the Order of Roraima of Guyana. The award is one of the two most prestigious that the government awards for outstanding service to the advancement of the nation and its people. Incidentally, we were so close that as she approached death, she gave me that medal, which I have today on the wall in my home.

On this particular visit, while I was in grad school at UC Berkeley and she was really aging, I was excitedly explaining to her what I was doing with these fancy experiments on the dynamics of the head disk interface and the tribological effects of landing an electromagnet using piezoelectric actuators on sheer layers of air adhered to spinning disks used to store information. Like a lot of young grad students at elite institutions, I was feeling especially smart and hadn't yet developed the maturity of translating what I know to what I understand. I was using dollar words for nickel thoughts. She listened patiently. She listened as a Black person born in 1916 listens calmly to one of her descendants born two generations later in a new world that is a descendant of the old. That listening is a kind of extraordinary combination of love and awe, patience and pride, caution and exalted wisdom above the details. She listened to my story against the long arc of her experience and rested me and my tales of excitement in her bosom of ages. She gave me anchor. Her parents, and their parents, were born toward the end of the nineteenth century. I have pictures of them on my wall.

In the British Empire, her grandparents were one generation removed from slavery. Slavery was abolished across the Empire in 1833 by decree, thirty-two years before American Confederates fought to the death to preserve it. She appreciated and chuckled at my unspoken sense of anointment and acceptance from a so-called elite institution. While I like to think I was immune to that insidious sentiment, I had to be vigilant against it, and my family helped ensure I didn't succumb to it. It is a form of approval and contrived self-importance from a system that only grudgingly grants you access in the first place. It is a false sense of arrival and self-worth. It is the framework of the idea that "I'm special Black people, not like the rest of them." I speak the Queen's English, not

Creolese. I speak corporate America—"Kudos to you," instead of "Say word." In today's American language, that is to be "acceptable to the white gaze." In Black-circle talk, when the white institution says you're the truth, then you must be the truth. Aunt Lynette had received the Order of the British Empire. For a Black subject of the British Crown, that sense of special anointed somebodyness could be overwhelming and destructive to her sense of connection to Black people. But her theme was that it was our cultural ties to Africa and the way we transformed them into our Caribbean identity that defined us, not a curtsy to the Crown. Behind her patience was the message: Don't be fooled. Don't be tricked into thinking that somehow all of this fancy training disconnects me from the people and story from which I come. It is neither acceptance nor approval. Merit is important, but it is not involved in the scorn born of racial animus. My presence at these elite institutions was tolerance and mere acknowledgment of accomplishment earned. Ours is not to seek approval, theirs is to seek forgiveness. Her message had always been that academic distinction was important, but that it doesn't make you distinct. Hers was a kind of full education, imparting to me the power of the bigger story and the longer arc within which to position what I know, so that I could understand.

When Guyana was granted independence from the British on May 26, 1966, Aunt Lynette was there as one of the captains of the new nation. She received a personal invitation from the royal secretary of Queen Elizabeth II and Prince Philip to join them on the HMY *Britannia* to mark the occasion. She has a picture of her curtsying to the Queen when they met—which I also have on my wall. When she discussed that with me, she spoke of the conflicting feelings. On one hand, the sense of honor and importance of being invited to meet the Queen, and the tremendous sense of the responsibility of independence, finally, of her country. Guyana had been colonized first by the Dutch in the late sixteenth century. Nearly four hundred years later, it was free. No longer British Guiana, but the Co-Operative Republic of Guyana. On the other hand, the staggering challenge of having to build and maintain a national and personal identity that was not defined and determined by generations of white British rule that had enslaved and subjugated us for centuries—a common challenge in the African diaspora. Were it not for the European enslavers, there would not have been Black people in Guyana in the first place. We had to forge anew. Her worldview spanned the range from learning folk songs with Amerindians, the native people of the region, to negotiating with the Queen about the future of the nation and being deeply tied to her African roots.

After she listened patiently to me talking smart about school, she said she wanted me to meet someone. We went to meet a carpenter she knew who had been in a terrible accident. He'd fallen off a ladder and had a catastrophic fracture of his arm. She forewarned me that his injury was gruesome and that we were not going to be voyeurs, but she wanted me to learn something about perspective. His forearm, between his wrist and his elbow, had been broken completely. He was wearing a leather Everlast weightlifting belt, the kind you would see in a gym. On the side of the belt, they'd affixed a platform so his arm could rest with his forearm parallel to the ground, his hand pointing forward with his elbow at ninety degrees. The repair was grizzly. His arm was screwed into the board. I could see the screws going straight through his skin, through his arm, into the board, and out the other side. It seemed incomprehensible. The kind of repair that, if I hadn't seen it myself, I wouldn't believe it as described. She introduced me to him and said, directly, that I was her favorite nephew visiting from the States and that she wanted me to see his arm and what happens to poor people in Guyana. He told us the story of what happened. It was a series of indignities tied to the ongoing national challenges faced by the country—a lack of plaster in the public hospital to cast his arm, a lack of foreign currency to pay a private hospital, the legality of private medical hospitals not accepting the currency of the country, and the unwillingness of the national health care system to subsidize the private system. It was an aggregation of small things tied to a newly independent Black nation facing enormous opposition in its attempt to develop.

As we drove away, she said that the advanced schooling I was receiving is important, but it is more important to her that I understand how to use these skills to make the conditions of our people better. She said that the British and the Dutch brought us to this land and sucked us dry and worked us to death for several centuries and enriched themselves on our labor and the resources of the land, sugar, bauxite, gold, and other minerals, and now we're left with a new country that can't repair a tradesman's broken arm. In that moment, she was less interested in whose fault it was than in my responsibility to see it. She wanted me to see the larger context and appreciate *why* this man had come to this horrible pass. While she didn't say it in so many words, her charge to me was that the purpose of education was to be able to see above the immediate circumstances and appreciate the historical context and variables that cause the conditions in which we live. Then, through the application of our learned skills, we orient what we do to the improvement of our people's condition. That was the whole goal of independence, to be able to exert not only *political* sovereignty but *intellectual* and *cultural* sovereignty—freedom. We needed the ability to determine, frame, and solve our own national aspirations and challenges. On reflection, she was pushing me to recognize the difference between education and training.

There is no question that both education and training are important. Overemphasis on the role and value of education, with its more intellectual goals of citizenship and sovereignty, opens the door to the obvious critique: You can't eat erudition. It doesn't pay the bills. You can theorize, but what can you do? Academics can certainly pontificate about the world in sometimes meaningless ways, using dollar words for nickel thoughts as a kind of shield from the "common person" who knows the reality of life but not the vocabulary of theory to describe it. Similarly, overemphasis on the trained trades is problematic. The argument that it is the trades that make the world move and provide for the solutions people need to live their daily lives and do the jobs to make a living is all true. The ability to manufacture, build, plumb, weld, code, and construct is invaluable to a functioning society. But too much emphasis there is its own risk. Aunt Lynette's point was exactly that. What I was prattling on about was mere training. Just as you cannot eat erudition, training doesn't necessarily help you see. How do you put the price of sugarcane into context, for example, and understand the crippling effects of generational subsidies from the European Union that are analogous to a form of national sharecropping when you're just trained to cut cane? Education helps put training into a framework. Oversimplifying the distinction between the two is helpful in a moment when AI is intruding and pushing and tipping the scale to training. Training masses of people, absent the context that informs the structure within which those trained skills sit runs the risk of fossilizing the structure that keeps them trained and trapped.

My doctoral studies brought me back to Guyana in 2003. I was interested in liberation as an objective of technological innovation. US and Western European scholars of national innovation systems at the time were nearly singularly focused on technological innovation as an engine of economic development. Simplified, their main argument was that the objective of national innovation systems—the confluence of public and private sector policy in education and investment to build a national science and technology enterprise—was purely economic. Understanding how nations or regions move from one category of development to another, and how the integration of public and private sector interests could enable that move, was the central theme of that body of work. In my reductionist interpretation, the self-interested goals for the US and European countries were dominance—determining how to lead the world in technological innovation as the next era of global dominance. To the extent that those studies applied to what is referred to as "the Global South," it was altruism—determining how to help the world's Black and Brown people help themselves a bit more, so they don't fall too far behind. It felt like a familiar theme. Needless to say, it is never lost on me that Western Europe and the United States were the primary coconspirators in the largest and longest execution of human slavery and colonial misery the modern world has ever known. They are the reason the Global South has the developmental challenges it does. For those European and American scholars, it seemed to me that their work was a study of the consequences of their own history of domination and subjugation. My interest was freedom—can newly independent former slave colonies acquire and assert scientific and technological innovation as a means of intellectual sovereignty? And if so, how? At the time, I was motivated by Rex Nettleford, a former pro-vice-chancellor of the University of the West Indies:

> Not least of our [Caribbean] crises is our continuing defiance of commonsense . . . which expresses itself in self-doubt, lack of self-confidence, continuing imitation of our betters so perceived, an intolerable intellectual indolence and a demeaning dependency. Help in this dependency is ever forthcoming from an outside world that has replaced military bombardment with cultural penetration and with the marvels of science and technology projected as the norm against which our ancestral wisdom is pitted as an aberration. Our small size and the absence of that science-and-technology knowledge-base, in any profound sense, adds to our vulnerability. Yet resilience and resistance are the *sine qua non* of human survival.[11]

The subsequent visit to Guyana brought me to the sugar estates. I was interested in the extent to which the sugar-producing states of the Caribbean have used technology to transform and take ownership of the industry or have maintained its long legacy of dependence on the Crown. The complex of international trade, economic forces, and the complications of being a former colony and member of the British Commonwealth are their own bodies of knowledge and study, which is where my work sat. The harvesting and production of sugar in Guyana, however, has been a continuous national economic engine since the early seventeenth century. I was fascinated by the fact that Black men in this century were cutting sugarcane on plantations in exactly the same way they had done as slaves under the Portuguese, Spanish, Dutch, and British for more than two hundred years.

On one visit to an estate, I went with a group of cutters out on their rounds. Being in the truck with these men was surreal and overwhelming. It was like a wrinkle in time, being transported back to another era. In the United States, the image of the slave picking cotton, for someone like me who grew up in Brooklyn, is an image of a distant time and another world. I'd never seen an actual cotton plant until I was an adult. The idea of what a person who was actually enslaved would have looked like, and my sense of the task of picking two hundred to three hundred pounds of cotton per day was all derived from photos, descriptions, and my imagination. My sense of pure cotton came from the cotton ball the doctor delicately puts on my arm after drawing blood. But there in the truck, I was with men who were doing exactly the same thing our forefathers had done under the discipline of the bullwhip. They were "first gang" workers going to do the very work that our enslaved ancestors had. I was sitting next to them and didn't need my imagination. They each had multiple machetes. The same long, heavy blade that has been used to cut cane by men like them for centuries. The edges of the blades were so sharp they gleamed. They needed multiple because they cut so much cane in a day that the blades grow dull. Each man would cut on the order of a metric ton of cane per day. That is more than two thousand pounds in the searing heat. And sugarcane is a type of grass. Its blades are as sharp as knife blades themselves. These men looked like they were woven out of chords of steel. And the smell of sweat—overwhelming. They were rail-thin, wiry, and pure muscle. Some looked as if their right arm was larger than their left. The veins in their arms were prominent and pulsing; they looked like snakes the size of straws wrapped around their forearms, slithering onto and around their fingers. When I shook some of their hands, they felt like stone. And these men were black. Their skin was the color black. Black Americans would say they were blue-black—so black they were blue. In the imagery of caste, color matters. Crudely, Black as a racial identifier has a broad range of actual color—from light-skinned to dark-skinned, a kind of ombre that positions you in the Black caste. The whole Black world knows the ranking associated with that range. It is the quiet and unspoken subtext to why many of the most prominent pioneering Black people in the United States who have representational responsibility over white people are typically light-skinned—Thurgood Marshall, Adam Clayton Powell, Barack Obama, Kamala Harris, Hakeem Jefferies, Cory Booker. Breaking the caste structure has to happen in increments. Black, yes, but light-skinned first. The country isn't yet ready for men who are cane-cutter black.

There in the truck, I was sitting with an unbroken human chain stretching back four centuries to the origins of the African slave trade. I am of these men; that was Aunt Lynette's admonishment. But for them and the mighty women who endured the unspeakable—and resisted and loved despite—there is no me. The violence and dehumanization, the need for separation of body and spirit, the yearning for God's redemption and salvation, the forging of love out of the ugliness of life, the refusal to break—it all surged. In Bob Marley's song "Give Thanks & Praises," he laments that our souls are hurting deep within and pleads for Jah to guide and protect us. I was seeing the source of that lament and wailing up close. It is a digression to share that I left the plantation in tears.

On reflection, this wasn't the kind of direct experiential education that a school can provide. But it was the kind of experience that illustrated, with powerful emotional clarity, the existence of a caste structure and how permanent it can be. These men had been doing the same thing for four hundred years. If not a permanent position at the bottom level of the caste structure, what is it? Nearly 20 percent of Guyana's population is tied to the sugar industry in some form.[12] So, in many ways, they can't stop doing what they're doing. Suggesting that old agricultural traditions are not valuable is not the point. This particular tradition is a result of the imposition of slavery. The inability to modernize the task, to make it more knowledge-based and technology-enabled, is a result, in part, of where it came from. It was not in the interest of the designer of the system—the British Crown—for the workers to move up, or for the nation to assert a parallel seat among the community of nations. The relationship to Britain for a brand-new nation meant ongoing dependence. The idea of equivalence between a former slave colony and the Crown is near heresy. The hierarchical relationship has been forged over four centuries. Guyanese independence was just six years before I was born. St. Kitts and Nevis, where my wife's family is from, and one of the six sugar-producing countries in the region at the time, earned independence in 1983 for the first time since the seventeenth century. I was in the sixth grade. Breaking the sense of hierarchy requires a new imagination and a jettisoning of the old-world structure. So far, independence has neither changed the hierarchy nor the country's dependence on other nations. You, former colony, can have your flag, but remain in your place. Oil is the next resource to be sucked from Guyana, but had that not been the case, what difference would the future hold for these cane-cutting men? Now that Americans have discovered oil, it is likely that these Guyanese men will lay down their machetes—finally, after several centuries—and simply pick up the oil wrench.[13] They'll be trained on the new skill but will go nowhere. They will acquire a new function in the same position.

Cheikh Anta Diop, a Senegalese historian and one of the preeminent scholars of African history in the twentieth century, outlined the point clearly in his book *Civilization or Barbarism*.

> The essential thing, for people, is to rediscover the thread that connects them to their most remote ancestral past. In the face of cultural aggression of all sorts, in the face of all disintegrating factors of the outside world, the most efficient cultural weapon with which a people can arm itself is this feeling of historical continuity.[14]

Understanding and finding place in the arc of historical continuity is one of the foundational goals of education. From the point of view of Black people facing aggression from the Crown or the Confederacy or modern populists looking to make America great again, that ancestral past is the wellspring of resilience. It is assurance that beauty is, in fact, deathless. It enables the process of training on new skills to be put into context with a broader and more courageous set of ambitions. The importance of that point of view is to be able to refashion the problem statement when it comes to new technologies, new training modules, and new promises of the future. Albeit just for a few impactful fleeting moments, my brief intimate touch with the long arms of slavery emphasized that point. As the world moves from one industrial revolution to another, it is absolutely essential to be wary of the variables that we hold as constant and those we deem mutable. The purveyors of new technologies may not, and often do not, have any interest in disrupting the hierarchy—especially the bottom rungs. They are technical, not social, revolutionaries. On the contrary, at best, they may simply be unaware of how what they're peddling affects the larger structural social order. They are, perhaps, so excited by AI in this instance that they fail to look critically at the potential reinforcement of the structure. At worst, they may have malicious intent. They may be looking to reify caste and turn it to stone by making the tools that keep people in their place so attractive that the entrapped carry the banner of their own entrapment.

Sitting next to men in the twenty-first century doing the slave work of the seventeenth reinforced the importance of education for me. Training can certainly exist without education, and education can exist without training. But a narrowly skilled, uneducated worker is at risk of facing generations of perennial hardship. Hence, some of the explanation for the permanence of cane cutters nearly two hundred years after the abolition of slavery in the British Empire. An educated worker, however, whether their skills be high or low, is dissatisfied with the seasonal repetition of oppression. Education, hope, and circumstance are the engines of that dissatisfaction. For sure, dissatisfaction is no guarantee of anything. But education is inescapably important for escape to be possible. The banning of DEI because it is a cornerstone of a complete education is an attempt to block the escape route. It matters little if the lens is broad and captures the African diaspora in the Caribbean—Georgetown, Kingston, Bridgetown, or Basseterre—or if the lens has a narrower US focus—Atlanta, Oakland, Detroit, or Brooklyn. Understanding historical context is the scaffolding upon which problem frames are built. What would happen if claims about the promises of AI in education confronted students in today's segregated school systems who were armed with James Baldwin's understanding in their curricular DNA and Cheikh Anta Diop's history in their cultural veins? In the face of these enthusiastic claims of what AI can do for education, what would the students say? First, they would focus their second-sight to see behind the facade. As Baldwin points out, "in order to save his life, he is forced to look beneath appearances, to take nothing for granted, to hear the meaning behind the words."[15]

Educated students would go behind the words. They would see the contradiction and cry foul. They might say that you cannot ban Advanced Placement African American Studies on one hand and offer AI certificates that are supposed to launch their economic mobility on the other. The two are incommensurate. They'd know they would be flying blind. They would understand the context of economic mobility and question the parameters that define racialized economic immobility. They would have learned about real sharecropping and positioned it as a historical reference against which to recognize modern digital sharecropping. They would be conscious of white people's historically enraged reaction to Black economic mobility, the very thing being promised, and be able to call on the ghosts of the Red Summer, the Tulsa Massacre, the Atlanta Race Riots. They would know, for example, that Atlanta is one of the most segregated cities in America and has one of the lowest rates of economic mobility for Black people and one of the highest levels of wealth inequality.[16] Given that understanding, they would be armed with the critical thinking necessary to ask if the AI certification training they're being offered is the right tool for the task. Armed with Baldwin's clarity of sight and Diop's presentation of history, they would ask who and what rules are responsible for the patterns of segregation and economic immobility that define their lives in the first place. They would be suspicious and wonder why you're offering certificates to them, but advanced computing at elite universities to your own children? Why is something supposedly good for them that isn't good enough for you? To whose benefit is it to have them in the workforce well trained but not knowing African American history—or American history, for that matter? They might ask, if equal opportunity as the bedrock of American democracy is condemned as illegal and immoral, does that mean that the inequitable education they receive is legal and moral? They would have a second-sight and not swallow the sedative as prescribed.

In 2023, Google partnered with Atlanta Public Schools to create the APS Grow with Google Career Certificate program.[17] The program was designed to offer Google certifications to students and the community. The certifications were in data analytics, IT support, project management, user experience design, digital marketing, and e-commerce. The announcement of the program declared that "Google Career Certificates prepare you for a new career in under six-months in high-growth fields, no experience required. You'll get professional training designed by Google." It went on to say that one of the advantages is you can "learn job-ready skills, even with no relevant experience and make your resume competitive with a credential from Google."

Corporate certifications are not new. They are part of a growing component of American education. In technical fields, the recent surge in certifications was coincident with the rise of Big Tech over the last fifteen to twenty years. All of the Big Tech firms offer some version of certifications that they endorse simply with the power of their names. Hence, you can make your résumé competitive by merely having a credential from Google on it. Their claims are invariably that you can grow your skills and career through certification with their courses taught by their firms' experienced technologists. Armed with their certification, some new career path opens—front-end developer, cloud practitioner, DevOps engineering, and the list goes on. In some ways, it was a reasonable and practical response to their growing demand for technical talent during the surge times of the early 2000s. At the time, computing and information technology programs at universities and colleges were simply not keeping up with the technical demands of these companies. The companies pushing the technical frontier were outpacing what universities were offering, so they decided to customize certifications to suit their needs and offer them themselves. It was a blurring of the boundary between postsecondary education and the private tech sector, with one taking on the perceived shortcomings of the other. With that came the slow creep of the shift from education to training à la carte skills that are singularly tailored to specific job tasks. There was also the reality of the tremendous growth in the sector. Using Google as an example, there were approximately 17,000 employees in 2007. In 2024, there were 183,000, a nearly tenfold increase.[18]

Of course, not all of those employees were CS majors, but it gives a sense of the expansion. By contrast, there were approximately 38,000 undergraduate CS degrees conferred in the United States in 2000. In 2019, there were 89,000, a little more than double.[19] The tech companies just needed a lot of people. For second-career workers and people on the career path in need of upskilling, these efforts were designed to expand the pool of people. It is a good thing to offer skills for meaningful employment and advancement built on the idea of continuous learning. The credibility behind all of it, however, is simply the power of the name of the companies. So awestruck are we by the power of the technologies emerging from the Big Tech firms that all we need to know is that Google certified the course, and therefore, it must be worth it. And given that it is taught by experienced Google engineers, or Meta engineers, or Amazon engineers, it can only be extraordinary. Waymo cars are driving around the city with no drivers; ergo, if the person who did that is teaching the certification class, let me be certified. In the case of the APS Grow with Google partnership, the language was explicit: "With no relevant experience [you can] make your resume competitive with a credential from Google." What was Google really telling the Black students of Atlanta Public Schools? That they could take an online course for a few weeks with no relevant experience, make their résumés competitive, and be prepared for new careers in high-growth fields? It is unclear whether students themselves bought it, and the data on if and how many students were able to make good on the promise of the competitiveness of their résumés and their preparation for new careers is hard to come by.

The unveiling of the program happened to great fanfare. The United States deputy secretary of education, Cindy Marten, came down from Washington, DC, to bear witness. Many of the local Google leaders were on hand to discuss the merit of the program. In corporate efforts like these, there is always a marketing agenda. When the primary KPI is corporate "wins," part of the goal is to be noticed, so the press was there. The language of these efforts has a particular cadence and vocabulary. Are the programs replicable and scalable? Can they have a tenfold impact? Can they mark the beginning of lasting movements? Can they leverage the expertise of the company to amplify upskilling in schools? In the background, there is also the quid pro quo of these efforts. Cities offer incentives for corporations to build headquarters or hubs in their municipalities, and these efforts of community engagement or corporate altruism are partly offered in return. Gestures like these "illustrate [fill in the company name]'s commitment to expanding educational opportunity for all students." While that commitment has very much changed since the change in administration, it doesn't disrupt the cadence of the process. It is the offering of certifications for a certain class of students commensurate with the expectations of where those students fit in the hierarchy. Atlanta Public Schools is nearly 80 percent Black students. If Google were to offer six-week certifications to middle-class white students with the assumption that it is an alternative to college, there would be revolt. But here, it is classified euphemistically as community outreach and engagement.

It is important not to be overly critical of well-intentioned efforts. The gadfly has its place, but too much is too much. Given the wave of AI certification and AI literacy commitments and tools that are flooding school districts now, however, it is necessary to pause and reflect on what is actually happening. These career certificates Google was offering to high school students in APS and their surrounding community came just before the arrival of OpenAI's ChatGPT, so they offer immediate historical perspective of how these partnerships ebb and flow. The proclamations of corporate-sponsored AI certifications to students and school districts are the same—a new technology with an old promise. The promises of AI to make teaching more efficient, more personalized, and more effective—that it will improve student outcomes—will abound. That is precisely what the Grow with Google Generative AI for Educators promissory note says, "As a teacher, we know your time is valuable and student needs are broad. With Generative AI for Educators, you'll learn how to use generative AI tools to help you save time on everyday tasks, personalize instruction, enhance lessons and activities in creative ways, and more."[20] That is the perennial language of the technological promise. Whatever it is that you do, we'll make it better. While there is an abundance of truth to that in some respects—with regard to education as an enforcer of caste structure—the promises run hollow. If what schools do is enforce caste structure and help undereducate Black students, we don't, in fact, want that system to be made more efficient. On the contrary, we want to see it disrupted. That is the core challenge of reframing the approach. What the schools are doing relative to Black students doesn't need incremental improvement in efficiency. Currently, they are efficiently choking the intellectual life out of Black students and crushing their ambition. The fact that their academic performance indicators decrease the longer they're in school is evidence of that. The cynic would say that making that process more efficient is merely speeding up their descent to the bottom of the well—making their peace accord with mediocrity easier to sign. Overlooking the relationship between the caste structure and the public school system in Atlanta, as was happening in this partnership between APS and Google, stains the certificate program as not only insincere but insidious.

The counterargument is that we need to meet students where they are. I heard this argument myself in discussion with some of the senior leaders of Grow with Google. College isn't for everyone. The cost of college no longer justifies the diminishing return on that investment. Students need to gain access to the workforce from wherever they are. Those arguments from senior leaders are often laced with personal anecdotes. The ones who make those arguments most forcefully are typically those who didn't follow the traditional path to the top. It often doesn't dawn on them that their exceptionalism doesn't translate. Carving a path to the top of a company like Google and not having had a traditional elite education is a rarity, even when you're white. Being Black, it is a virtual impossibility. The argument doesn't transfer. The logic doesn't hold when suggesting that Black students can wrangle a six-week certification and then do something meaningful like launch a new career in a high-growth tech field. At best, it is faulty logic; at worst, it is a sham. It is lost that while these claims for certification as a path to economic mobility and the launching of new careers were taking place, rich white families were being arrested for cheating and scamming to have their children attend the nation's most elite institutions. The Varsity Blues scandal saw rich and famous white celebrities and business leaders go to jail for cheating to get their children into elite schools.[21] *Why*, when the certificates were free? The relationship isn't one-to-one. My Google colleagues were not involved in that criminality, but the larger point is one of hypocrisy. To the Black students, we offer a six-week certification upon which to build a new trajectory for their lives. But rich white families are willing to lie, cheat, and deceive to get their children into the institutions that actually do launch careers and make résumés competitive. Here, too, the price of the ticket is an indicator of value. A computing degree from Stanford makes your résumé competitive, as it rests on the sum of the most advanced undergraduate technical education in the world. A six-week online certificate for a high school student does not. The recipient of the former is the one who builds the certificate program for the latter. They then, with their children securely on the legacy path back to Stanford, Carnegie Mellon, UC Berkeley, or MIT, turn around and offer the certificate that they built and endorsed to the Black kid in Atlanta Public Schools as a path to in-demand careers and economic mobility. The good intentions behind the effort are not a mask for what it is. We are offering Black students training to be the IT help. It may be in information technology, but they are still the help. It is a paved path to a technical ghetto. Yes, being in the tech sector at any level is likely an economic step up from concentrated and generational poverty. There is no denying the benefit of programs like these in that regard. The problem is they don't do anything to change position. It is merely a change of function. I don't believe that there was malicious intent among the people who designed this program. Some of them I know. The problem is, as in Isabel Wilkerson's caste assessment, they just didn't see the invisible ruts determining the a priori outcome of their effort. Let me help you stay behind, just not that far.

According to the State of Education in Metro Atlanta Annual Report 2024, the high school graduation rate in metro Atlanta area school districts in 2023 was 84 percent.[22] In 2024, the high school graduation rate in Atlanta Public Schools—the beneficiary of the Grow with Google Certificates—was 88.4 percent. So proud was the district that in the press release, they didn't round the .4 percent down to 88 percent, they included it to indicate that the four-tenths matter—88.4 percent. Each tenth of a percent represents more students and how much better APS was than metro Atlanta. That data came out with praise and accolades. It was heralded as the "highest all-time graduation rate for the district." The notice went on to say that the APS graduation rate had surpassed the state's graduation rate of 85.4 percent by 3 percentage points. Accompanying the announcement was the proclamation of the superintendent, "it is our intention to continue building upon what we've accomplished to ensure APS produces high-performing graduates who are prepared for a four-year university or technical college, ready to obtain a desired certification or credential leading directly to work, and a future that guarantees a successful career and life." Here, too, the superintendent is a good man and is absolutely committed. In his proclamation, he covered all the options—university and technical two-year institutions, as well as direct-to-work certifications and credentialing. Under that cover, the only things the district doesn't support are dropouts and graduates ill-prepared for the modern economy. But what is this graduation rate really saying? The graduation rate is measured by the number of students who graduate high school in the four-year span allocated for that phase of secondary education. It is an objective calculation. Eighty-eight out of every hundred ninth graders enrolled in an APS high school graduated from the twelfth grade four years later. The praise that rained down on the 88.4 percent suggests that it is high-caliber teaching and instructional and structural improvements that led to that record-breaking rate. The suggestion that the district needs to build on that accomplishment is an affirmation of that implicit assumption. We're doing it right. There is surely the Herculean effort by hundreds of fully committed educators, administrators, parents, and students themselves making that happen. But the other way to improve a graduation rate is to change what is required to graduate. In addition to data about high school graduation rates, the State of Education Report offers data about the five-year college graduation rate. It offers a picture of what happens to those 88.4 percent of students who graduated from high school. The five-year college graduation rate in 2024 was 26 percent. Only a quarter of the students who graduated in or around the time of the highest high school graduation rate in history graduated from college. That is an inconvenient contradiction. Sadly, it is an indictment of the utility of the high school diploma. Clearly, the criteria used to assess the high school graduation rate are not the measures that make students "college ready," which is the first clause of the motto of Atlanta Public Schools: "Making Students College and Career Ready." If proficiency in basic algebra were one of the criteria for high school graduation, only 11 percent of Black students in Atlanta Public Schools would graduate, and only approximately 14 percent would graduate in America.

The argument that Google and other tech firms peddling their training certifications make is that they are meeting students where they are. They are offering students a foothold in a swiftly changing occupational environment by providing certification of entry-level skills. That argument is sound and ought not be dismissed. The question, however, is whether it is ethically sound to offer the drowning person a lifejacket but refuse to pull them onto the boat. The answer to that may not be so straightforward. What is certain is that, armed only with a lifejacket while adrift in the water, their lives are still defined by the struggle to survive. The luxury of rest and reflection offered by a seat on the boat is not afforded to them. Yes, there is a case to be made about the imagination, ingenuity, and courage forged in the crucible of pressure, pain, and the desperation to survive. That is the case for agency—the ability to take the reins of your own life, however prescribed, and change it. If all you have is lemons, make lemonade. That is the bootstrap case. Yes. Black people know it in our DNA. Our ability to survive when America was allegedly great is rooted in that strength and fortitude. But that is not a contemporary strategy for the upliftment of a people. Pain and suffering and the proximity of death ever-lurking ought not be the systemic motivation for structural innovation. That is a lot to ask of a generation born in the twenty-first century. Having agency and a sense of self-determination in an educational system that is fighting against you—stripping education and offering menial training—is a rigged fight. The apparent permanence of the Black underclass in America is evidence that the pressure strategy doesn't work. That is the romantic overemphasis on the exception; that is the overlooking of the pattern. If masses of Black people are fighting just to survive, if we win, "survive" is likely all we will do. Translated, that is the goal of the bottom caste—survive.

That is how the Grow with Google Certificates work together with the yawning gap between the high school graduation rates and the paltry college graduation rates. The data are irrefutable that the high school graduates are coming out of school ill-equipped for college, and "career" needs a definition. The offer of certification is the lifejacket. A desperate school district will gladly accept the offer to help students they know are at risk of drowning. Train them, lest they drown. It is a simple calculus. Teachers in the trenches in deeply segregated Black schools, gasping for air, dealing with students day-to-day, know intimately what their students are capable of in ways that no one else does. When asked, they will talk of the towering ambition and creativity and even optimism of their students, but also how those traits are undermined by the nearly criminal underdevelopment of skills and structural constraints on their education. Faced with the moral conflict of demanding a seat on the boat or preventing the students currently in the water from drowning, school leaders make the obvious choice. No individual student is sacrificial. Therefore, they accept the lifejacket—the certificates that may or may not lead anywhere—and forgo a seat on the boat as an ask too far, a demand beyond our station. Hence, our people end up with upgraded training for their lot in the same spot: the bottom.

It is relevant to note that while the promises of advancement for Black students were being offered based on generative AI and other training certifications, Google settled a $50 million class action lawsuit for discrimination against Black people, the largest in its history.[23] Like the high school graduation rate, it was another historic first. The coincidence of these two realities affirms the significance of training in a system that is reinforcing hierarchy. It exposes the expectation behind the offer, whether spoken or not. If it is assumed that Black people will be on the bottom, then the training offered to them is commensurate with that assumption, and the expectations of the outcomes are preordained. The expectations are that Black people will be on the bottom. Those expectations are demonstrated not only in the training offered but also in the treatment afforded—underpaid, under-leveled, highest proportion of lowest performance ratings, least likely to be promoted, longest time to promotion, first to be laid off, highest rate of mental distress, highest rate of job dissatisfaction, lowest rate of sense of belonging, et cetera. Nothing new in the American story. Yes, you can be trained in high-growth careers and have competitive Google-endorsed résumés, but you will still face the weight of the caste anvil bearing down on you. And what does it mean to the Black person to have their résumé endorsed by the company that discriminates against them? In some ways, it is a trap—accept the lifejacket lest you drown, but be relegated to a life dragging behind the boat.

There is no easy path out of this conundrum. We have become accustomed to the three-, five-, or seven-step simple guides to some desired outcome. This isn't that kind of problem. This is a problem about what we collectively know and understand about this country and the role education plays in achieving that understanding. In the simplified view of the distinction between education and training, it is easy to bend to the argument of pragmatism. People need skills to work, and they need to work to live. Therefore, the need for training to acquire skills always comes out on top. It is a survival argument. In the contemporary political fray about these issues, the defense of training is trumpeted as "you can't eat democracy"—in other words, that we need to focus on the economy and kitchen-table matters. That is the easy prioritization of training. It is also the convenient slot for AI literacy in schools to fall into. Rather than defend the essential role of education in developing a complete national understanding of who and what we are as a country, we can hide from the assault on DEI and forsake education to appease its rabid attackers. AI literacy as a form of contemporary training can serve that role, and it is. It can serve as an appeasement. At some level, it is merely training masquerading as collective investment in transformative educational technology to improve mass upskilling. That is not education. It is true you can't eat democracy, but it is also true that you can't eat tyranny. And it is the uneducated society, whether trained or not, that is most susceptible to the whims of a tyrant. History is overflowing with examples. While the dominant narrative about the United States is that it is somehow above such third-world follies, it is not. And Black people know it. Our education was, in fact, forged in the pressure crucible of pain and suffering. The legacies of Ruby Bridges and James Meredith and the Little Rock Nine attest to how valuable education is. Black people at that time were willing to die for it, and white people were willing to kill to prevent us from getting it. Knowing that at all is a result of American education, not training. It isn't coincidence that, today, the group of American people who perhaps know this best are the ones whose education is being attacked first. According to one of America's greatest heroes, Frederick Douglass, "It is easier to build strong children than it is to repair broken men." At risk of rhetorical hyperbole, there is no other way to build a strong democratic nation than through the development of an educated citizenry. The educated, valued, and respected citizen armed with rights and education on the history and protection thereof is the cornerstone of such a society. This is a critical time in the country. The doubling down on racial hierarchy and animus is unlike anything we've seen since the end of the first Reconstruction. The surge of technological advancement is equally without precedent. Given that, it is critical to see the transference of the theme. It is easier to build a strong nation than to repair a broken one. National strength comes from education.

PART IV

MUST BE FREEDOM

CHAPTER 7

MODERN FREEDOM OR DIGITAL SLAVERY

Artificial intelligence has introduced enormous social and philosophical questions—staggering ones, in fact. We're grappling with words and arguments that are hard to comprehend and translate into the daily rhythm of our lives. To date, these dystopian concepts have felt more like the stuff of novels and movies than real life. Existential questions about the future of human agency—whether fatalistic or optimistic—abound. Will AI replace you? Will human intelligence be distinguishable from artificial intelligence? What does the future of human agency even mean? What is the current state of human agency against which the future state will be measured? Does artificial general intelligence mean that with some flash of light, in a magic moment of singularity, intelligence will be transferred from the human mind to models deep within the digital teraverse and leave us in the sunken place? Where does that leave the soul, and whose side will God be on? Will American race ghosts be exorcised, finally? All fantastic stuff. But this whole enterprise is the product of people. Therefore, it is subject to the vices and demons of people. Given the grandeur of the potential of artificial intelligence, scholars and technologists claim that it is catalyzing the Fourth Industrial Revolution—a fusion of digital and cyber-physical systems. The technological advance we're experiencing with AI is deserving of the hyperbole. But so is the resurgence of racial vengeance and animus, and America is chained to its history of free and slave, of white and Black. That tension, unresolved still, is raging in front of us again, now complicated by mind-bending technology. The two are connected. To think otherwise would be naïve. The potential power of these new technologies brings with it the dual sensations of awe and fear. When fear rears its head across American society, it conjures the ghosts. History suggests caution that we not be seduced by technology and expose ourselves to the shape-shifting ghosts of the free and the slave.

If we follow the arc of industrial revolutions, it takes us back in time and offers some perspective. The first industrial revolution in the late eighteenth century saw the mechanization of formerly manual tasks. Eli Whitney's invention of the cotton gin in 1793 is the most famous product of that era. It mechanized the separation of seeds from the valuable cotton fibers and made cotton the undisputed king. In today's language, it increased the throughput of a critical bottleneck in the supply chain. As a result, it exploded the demand for slaves in the American South. The number of enslaved Black people rose from seven hundred thousand in the 1790s to nearly four million by 1850.[1] The Second Industrial Revolution, the Technological Revolution, in the late nineteenth and early twentieth centuries saw the mass provision of electricity, the production of steel, the laying of the railroad system, and industrial-scale manufacturing. It also gave rise to the robber baron class. It came with the amassing and concentration of fantastic wealth in what would be known as the Gilded Age of the Carnegies, the Rockefellers, and the Vanderbilts. It set the stage for the Great Depression and coincided with the rise of the Ku Klux Klan. In the early twentieth century, at the height of the Second Industrial Revolution, there were an estimated four million members of the Klan,[2] marking a uniquely American cultural milestone. The Third Industrial Revolution began in the latter half of the twentieth century and was marked by the early shift from analog to digital technologies. The development of personal computers and the birth of the internet were the key markers of this age. The technical titan of the second half of the twentieth century was IBM—Big Blue was at its height in the 1960s through the 1980s. Microsoft was founded in 1975, and Apple in 1976. The Third Industrial Revolution coincided with the long tail of violent opposition to the Civil Rights Movement. It coincided with raging resistance to school desegregation, the beginning of the crack era, and the mass incarceration of Black people. In 1970, the prison population in the United States was approximately two hundred thousand. By 2009, when Big Tech was exploding, the prison population was approaching two million people, nearly half of whom were Black.[3] As the Big Tech enterprise was gaining steam and fueling the engines for the launch of the Fourth Industrial Revolution, Black people were being sent to prison.

Now, here we are at the dawn of the Fourth Industrial Revolution, in the undisputed era of artificial intelligence. It is surging at the end of the first quarter of the twenty-first century and overlaps with the 250th anniversary of the country. Like its industrial revolutionary ancestors, it is coincident with a rise of racial animus. The gutting of American public education, the branding of Black studies as radicalized terrorism, unmitigated racial hostility, and instigation from the president of the United States have returned. The demons are back. Each of these technical revolutionary periods was marked by superlatives about their transcendent effects on the world. Yes, their effects were profound. But American ghosts remain undeterred and unpersuaded. None of these revolutions had any effect on the sentiment so perceived. None of the claims—the alleged improvement on the quality of life for everyone, or increased worker productivity, or improved efficiency of innumerable tasks from teaching math to manufacturing—have done anything to assuage the ghosts.

While each revolutionary era may have been wrapped in platitudes very similar to those we're experiencing now in the Fourth Industrial Revolution, history gives us clarity of sight and the ability to see behind the words. Just as the ubiquity of technology in our lives didn't prevent officers of the state from strangling George Floyd to death, neither will AI prevent racial animus from strangling the country. On the contrary, it may tighten its grip. The workforce development argument is convenient and nice sounding, but it is not up to the task. The president issued the order to destroy the moniker that said Black Lives Matter, and he issued the order to advance AI in education. The two are not necessarily contradictions. A more technically trained workforce, stripped of a whole American education, wasn't there to stop the destruction from happening. Technical advancement is exciting, but it does not lead to the exorcism of America's ghosts. Mass deployment of breezy no-experience-needed tech and AI certification training, for example, looked at from this point of view, is its own version of sharecropping. It is its own version of contract labor—in fact, it *is* contract labor. It fits the pattern.

Technological progress does not yield social progress on its own. The application of a technical solution, however awe-inspiring the technology may be, to a social problem does not guarantee anything. There is no causality there. Technology didn't cause the problem, so we cannot expect it to solve it. The cotton gin didn't create slavery, but it did provide the technical thrust to make it much worse. The massive expansion of the railroads was an enormous socio-technical leap forward for the country. It was a cornerstone of the Second Industrial Revolution. The argument that it would provide connectivity and opportunity and serve as the steel veins of American productivity and prosperity was true. It also produced an opportunity for American ghosts to take advantage of the loophole in the Thirteenth Amendment, which allows for the enslavement of convicted felons—"Neither slavery nor involuntary servitude, except as a punishment for crime whereof the party shall have been duly convicted, shall exist within the United States, or any place subject to their jurisdiction." It provided for the profiteering of prison contract labor of Black men. It was slavery by another name.[4] The claims of how AI in education will improve learning and close gaps and make students more prepared to take part in the modern workforce are colliding with historical patterns—ironically, the very history that is being summarily banned under the banner of illegal and immoral DEI.

It is during the technological leaps forward that claims and warnings proliferate. In the AI education world, the claims appear to be both grand and ahistoric. The benefits to education, and to rapidly advancing the skilling and schooling of Black students and the districts in which they sit, are ringing out from coast to coast. The reality, however, may be that American demons are conjured especially during phases of big technological leaps forward precisely to ensure that Black people do *not* participate. Participation at the frontier of technological advancement at these critical paradigm shifts comes with the ability to shape the contours of the country, to construct its future, and amass fantastic wealth. It is a form of authorship and captaincy—the sole source of power of the Department of Government Efficiency, for example. It is the ability to migrate from the margins into the main text and author the American story. Coauthorship requires a fundamental acceptance of racial equality. There is no alternative to that. If we are to coexist, we have to be coauthors—and our two versions of the American story must cohere. There must be, finally, equivalence and reconciliation between the bullwhip and the tattoo.

Modern freedom or digital slavery—hyperbolic language? Perhaps, perhaps not. My father, Robert Bobb, was from Guyana in South America. He was a Pan-Africanist in his blood and deeply committed to Guyanese independence in his youth. He felt that one of the most important goals for Black people of my generation was to shake the shackles of the long chains of slavery and colonial dominion—not just over our lives but over our minds. He chanted along with Bob Marley and the Wailers, that we "have to emancipate ourselves from mental slavery." Even my name, Kamau, is Kikuyu. He was inspired by the Mau Mau rebellion led by the Kikuyu against the British in Kenya. Mau, he would say, "we as a people have to take the Oath and fight for our freedom." The first Kenyan president, Jomo Kenyatta, was a towering figure of the Pan-Africanist movement, and one of his family names was Kamau. Hence, I am Kamau.

My dad taught math and physics and, like a lot of Caribbean immigrants to the US of his generation, felt that the American education system was woefully inadequate. My elementary school in Brooklyn was full of students like me, either first-generation West Indians or students born in Guyana, Jamaica, Trinidad, St. Lucia, Grenada, or St. Vincent. He offered after-school lessons to me and other Caribbean students in my class to keep us at the level he thought we ought to be. Those lessons were built around mathematics and Black liberation. It was in those lessons, in fact, where I was introduced to what a hyperbola is and, therefore, what hyperbolic means—both in math and in rhetoric. He insisted that we understand the origins of words as the first step in the ability to see behind them.

When he was teaching us the fundamentals of geometry, he introduced the Cartesian plane and the coordinates that specify points and serve as the visual map of Euclidean axioms. His analogy, I'll never forget. The origin of the Cartesian plane is the center of the entire coordinate system. It is the point defined by the coordinates (0,0). It's the intersection of the x- and y-axes and the point against which all others are defined (x, y). In his lesson to us, the way the long arms of slavery and colonization continually entrap our minds is by defining the origin. Europeans generally, and here in America, certainly, white people are the point of origin. They are the group against which our position in the world is measured, and more insidiously, our value is correlated with proximity to them. The closer we are in form, function, and identity to white people, the more valuable we are as people. That was the trap. He laid out the challenges associated with colorism; the schools we choose to attend; the way we talk; how we view our hair—nappy or pretty; the kind of names we choose—Jelani or Liam; the value we ascribe to our own institutions—Spelman or Barnard; the ability to "pass" as a means to escape the lynching tree. His point was that Black people are in a complex flux of yanking our mental and physical independence from a trap that has been laid out for us over centuries. One side of that is dependent on our own agency and the wherewithal to seize sovereignty, to emancipate ourselves from mental slavery. But we're not in the flux alone.

It is the reconciliation between 1619 and 1776. The core thesis of the Make America Great Again movement is the centrality of white people as the origin of the Cartesian plane. When the heroes in the story are the white men who killed Nat Turner, they are the origin of the coordinate plane. The obsequious Uncle Tom from the Space Traders story thought that his proximity to white people would save him; it did not. Latinos who voted for this skewed American thesis thinking that they followed the legal path and that that would save them overlooked the fact that they, too, are distant from the origin. Their status as non-Black and of Latino origin, and even perhaps being white in the countries they come from, isn't enough to save them from El Salvador. That basic construct, white Christian nationalism and the centrality of white people as the origin point of American identity, has to be ceded for the national experiment to work. That is not a Black task. White Americans must do that. It will not be undone overnight, and it certainly cannot be undone without discernment and the exorcism of ghosts. The resistance to that is what is happening here in the background behind all the excitement about AI. Enthusiasm for AI is not a substitute for that national assignment. AI cannot be used as an excuse not to apologize. It is not a rock we can hide behind.

On reflection, the significance of my father's geometry lesson is that the language of freedom and slavery is not hyperbolic for our time. In this season of artificial intelligence and technical paradigm shift, the gravity of what it means to achieve modern freedom and escape digital slavery is very real. Technological progress and justice are not the same thing. Through his absolute reverence for, and insistence on, complete education, and on a Pan-Africanist world view, I learned about Hector Pieterson. He was one of the first school children shot by white South Africans in 1976 when they insisted that education be offered only in English and Afrikaans. It was an attempt to erase and diminish Zulu and Xhosa, the dominant languages that Black South Africans speak. It was the Afrikaners' version of banning DEI, of banning AP African American Studies. It was their attempt to reify white South Africans and their language as the center of the coordinate plane in Black South Africa. Black students, school-aged children, ten-year-olds, and fifteen-year-olds refused the indignity and revolted in Soweto. The Afrikaners shot them dead for daring to shift the origin. They killed hundreds of Black children that day.[5] Those white South Africans, the authors of Apartheid, dripping with bloodthirst, are now the very ones being granted asylum in the United States. They are of a kind with the killers of Nat Turner and Emmett Till, Steve Biko and Chris Hani, and the British orchestraters of the mass slaughter of Black people in response to the Mau Mau rebellion.[6] Attempts to shift the origin have been met with murderous bloodletting, and those killers are heroes in the current American Cartesian plane. We are in an era when this kind of shocking bloodthirst is not the stuff of history books or legend. It is in the air we currently breathe. Wholesale human slaughter in the Middle East, in Sudan, and in Eastern Europe, and the snatching and kidnapping of students and immigrants from classrooms, from off the streets, and from their homes in the United States today, is the forebreeze of the violence and terror approaching. It should have the hairs on the back of our necks standing. It is the precursor to an age we've seen before in America—when trees bear strange fruit, and Black communities burn for shifting the origin to themselves. Tulsa. It is lurking ever near. This bloodthirst is encroaching, like vampires amassing at the gate, threatening to burn the veil. Freedom and slavery, life and death, are not hyperbole.

In this environment, the promise and threat of AI only serve to amplify the precarity of the moment. What is certain is that Black people are systematically locked out of the seats of power associated with its creation. In 2013, the Center for Education and the Workforce at Georgetown University published a report, *Separate & Unequal: How Higher Education Reinforces the Intergenerational Reproduction of White Racial Privilege.*[7] It offered that over the fifteen-year period between 1995 and 2009, the percentage of new Black students enrolled in higher education increased by 73 percent, and the percentage of new white students enrolled increased by 15 percent. Both good signs. More American students going to college is positive. But of the newly enrolled white students, 82 percent enrolled in highly selective institutions, while only 9 percent of new Black students enrolled in highly selective schools. The vast majority of Black students enrolled in open-access and two-year institutions where the completion rates are notoriously low, resources are scant, and the prospect of education for AI, as opposed to AI in education, is slim. And that was over the roughly two decades leading up to the overturn of affirmative action in the *Students for Fair Admissions* case in 2023. The already limited access to American top schools, where the educational and research base for AI is gathering steam, is now almost completely closed. That restricted access on the education side leads to restricted access at the frontier of the private sector tech ecosystem. It is an unvirtuous cycle that renders Black people locked out.

The tech titans, none of whom were Black, stood shoulder to shoulder with the president and tacitly aligned with his explicit denigration of Black people and skewed view of the future of the country. They immediately dropped whatever tepid commitments to fairness and justice they had. It was such a pivotal point for the future of the nation. They could have made a different choice. As captains of the AI technological leap forward, they are the locus of tremendous power, perhaps unprecedented in modern American history. They could have stood together in defense of the promise of the United States. It could have been a monumental break in the pattern, the arrival of a new American day when the conjured ghosts were pushed back. Instead, they chose to build a new Death Star while the country brings back the slaughterhouse. History is riddled with wistful longings for could-haves. Survivors of the lynching tree and the cane field, the internment camps, and the concentration camps all know that things could have been different. And we recognize pivotal moments like these, where choice is presented, and courage is required. For now, disappointment rules the day, and history will have to judge. The net result of this is that Black people are being blocked from the national captaincy, blocked from authorship in the development of AI as the engine of the Fourth Industrial Revolution. The pattern of the technological paradigm shift, the reinforcement of the bullwhip as the American origin point, the exclusion of Black people, and the surge of racial violence is revealed again. Education is being whitewashed to wash out black and pave a path to digital slavery. Just as at the end of the first Reconstruction, slavery in

America name-shifts and shape-shifts, but it is ever-present. Behind the claims of AI in education and training, the constraints on access to elite technical higher education, and the collapse of courage among the leaders of the world's most powerful technology, James Baldwin's prophetic message rings out, "in the attempt to correct so many generations of bad faith and cruelty . . . you will meet the most fantastic, the most brutal, and the most determined resistance."[8]

This is that time in the United States—a time of brutal and determined resistance. It is a resistance to a shift in the origin, to a tearing of the caste structure, to an attempt to correct so many generations of bad faith and cruelty. We are in the throes of a collision between a historic technological leap forward and a dangerous social lurch backward. We're being seduced by AI to dream a new world, while we're being summoned by the bullwhip to reclaim the old. There is no easy way out of this tug of war. The centrality of AI in education, and in the conduct of nearly all business and commerce, makes it inescapable that we need to be AI literate, to learn and teach, and to deploy training and proficiency. Yes, we have to be pragmatic. We also have to protect and enrich education in its totality, such that not only can the nation know its true self, but we can create and seize the kind of education necessary for Black people to participate and lead in this AI-fueled Fourth Revolution. Six-week training with no experience, we can surely agree, is not the path. Full education is also necessary to build the guardrails, so the Death Star is not without restraint. The nefarious uses of AI-enabled technology for making killing more efficient, for making surveillance more efficient, for making the invasion of privacy more efficient, for making sexual predation more efficient, and for making racism more efficient are all just as real as the claims for making teaching more efficient—and the nefarious uses are all more profitable and likely more popular. Of course, we don't see billboards advertising AI for surveillance—find the most efficient means of rounding up twelve million immigrants and sending them to the terror camps. Nor do we see billboards advertising AI for replacement—find the most efficient means of total structural starvation and genocidal removal of 2.5 million people. We don't see billboards advertising AI for the undereducation of Black people—find the most efficient means of systematically keeping Black people out of the intellectual core of the most advanced technologies. There aren't any glossy, warm-colored ads with beautiful smiling racially ambiguous people for these uses. But being unspoken doesn't make them untrue. When the sentiment of the bullwhip is lurking, guardrails are even more important.

The challenge we now face is not just about the future of Black people staring down the threat of digital slavery. It is a national challenge to us all to define modern freedom in modern America. The white nationalists have adopted the term "our country." They surely don't include me in their possessive pronoun. When they declare their intent to take the country back, take it back from whom? In their view of their country, Black people as digital slaves is a desired outcome. That is the 1776 view of America. The 1619 view offers our second-sight. The blasting of our flesh to carve the tattoo offers remarkably clear vision. It offers the moral high ground and the opportunity to forgive. The mere fact that Black people are not consumed with vengeance is evidence of our willingness to forgive. If the bullwhip would extend with contrition, we could in fact be American equals—not just Americans. Our challenge is to envision a new path. Clearly, precedent does not serve as prescription. This fight from a kind of white America to restore and enshrine its position at the origin of the Cartesian plane has nearly obliterated the federal government of the United States. It is undoing higher education as we know it. It is dismantling our research capacity and our ability to know and understand. It is blinding us. It is pitting American citizen against American resident. It is glaringly sacrificing competence and merit for fealty. It is fomenting violence and crudeness, providing a welcome substrate for the kind of depraved bloodletting going on around the world. Their willingness to destroy and impart such widespread suffering and carnage is remarkable. Their commitment to refusal, astonishing. Surely, that is not modern freedom.

Our challenge is that the United States has no precedent for modern freedom. There has never been a time when America was great for everyone. It will require all our commitment, courage, and conviction to envision something we've neither seen nor experienced. In the world of AI and education, that clarity of sight and requirement of faith in something we've not seen holds true too. It is possible for the tremendous power of AI to be equitably applied to education, but only if anchored in justice, the cornerstone of freedom. That possibility demands we be honest with our intentions and reconsider the relationship between our assumptions and expectations. To date, those have been fossilized by the American relationship between the bullwhip and the tattoo. The current cleaving of the country clearly exposes that truth. It also affords us the opportunity to reframe the problem statement. In any effort of AI in education to improve, to train, to develop, to educate, we must ask, does it render the path to freedom or slavery? In each instance of what we're doing, the question must be, *will this render freedom*, and for whom? That is the calibration appropriate for this time of technological paradigm shift and the resurgence of ghosts. Establishing and maintaining an anchor to freedom is required if we are to see our way through this confluence of AI, hallucination, false neutrality, and the raging battle over the position of the origin. If we plan to escape a season of digital slavery for the coming quarters of this century, we have to be explicit about freedom. The ghosts of Jim Crow, of Apartheid, of the Nakba, and of Kristallnacht are hovering, hiding behind the blinding light of these new technologies. Our commitment to the bigger ideal is demanded of us to keep them at bay and exorcise them, finally.

In addition to teaching me math, my father insisted that I think deeply. He absolutely demanded it of me. It was the cornerstone of our relationship, and I loved him for it—in ways for which words are insufficient. He required that I not accept anything at face value, particularly if offered by a system designed to subjugate me—"the Babylon system is the vampire," was another of his Bob Marley chants. See behind the words; look beyond the veil; develop a second-sight; boy, use your coconut—those were my dad's incantations. When I was young, he banned me from watching shows like *Tarzan*, *Good Times*, *Sanford and Son*, and *The Jeffersons*. He felt they were attacking my subconscious sense of who was sitting at the origin point. So protective of my mind was he that he caught me watching *Tarzan* when I was in elementary school and threw the television away. For years, we had no TV. I had to read about Black people's experiences in the world. He made me read James Baldwin's *The Fire Next Time* when I was in the fifth grade. I had to read and reread and discuss and learn the vocabulary and concepts with him, especially the particular paragraph about the Negro's past:

> This past, the Negro's past, of rope, fire, torture, castration, infanticide, rape; death and humiliation; fear by day and night, fear as deep as the marrow of the bone; doubt that he was worthy of life, since everyone

> around him denied it; sorrow for his women, for his kinfolk, for his children, who needed his protection, and whom he could not protect; rage, hatred, and murder, hatred for white men so deep that it often turned against him and his own, and made all love, all trust, all joy impossible—this past, this endless struggle to achieve and reveal and confirm a human identity, human authority, yet contains, for all its horror, something very beautiful. I do not mean to be sentimental about suffering—enough is as good as a feast—but people who cannot suffer can never grow up, can never discover who they are. That man who is forced each day to snatch his manhood, his identity, out of the fire of human cruelty that rages to destroy it knows, if he survives his effort, and even if he does not survive it, something about himself and human life that no school on earth—and indeed no church—can teach. He achieves his own authority, and that is unshakable.[9]

He was a teacher, yes. But he was *my* teacher, and I love him in my cells. I am of him. He died while I was writing this book, so he won't have the chance to fully see how deeply I've learned his lessons and how well he prepared me for the battle ahead. I pray that he knows. It is an unconscionable ache. I had no idea how painful grief could be. I understand wailing now. The intensity of my grief must be correlated with the thickness of my love and how deeply I felt he loved me. I can't comprehend it any other way. My sense of loss derives from the power of his lesson. It stops me from breathing when I try to grapple with it. But his voice and demands of me were preparation for our time. He foresaw what was coming, and his instruction was to "keep the faith." He demanded critical thinking not just of me but of the thousands of students he taught over the arc of his life.

When he taught us as children about the Cartesian coordinate plane, we couldn't even take that at face value. We were expected to ask *why* it was called the Cartesian plane. He introduced us to René Descartes, the seventeenth-century French mathematician who revealed the relationship between geometry and algebra. Another incantation of my father's that governed my life and informs this critical season of artificial intelligence came from Descartes' *Discourse on Method and the Meditations*—"*cogito, ergo sum*," I think, therefore I am.[10]

So now, what am I to be if not to be free?

ACKNOWLEDGMENTS

My wife, Lisa, and our daughter, Sadira, provide the triangulated love that enables anything meaningful in my life to happen—Lisa, through the long arc of love, and Sadira, my anchor for hope in the future. This book is evidence of the convergence. The arguments I make have, in phases, depressed me, enraged me, and inspired me. Lisa and Sadira helped me remain steady in the storm, with clarity of sight and precision of words throughout.

I wrote this book in a nighttime of my life. I lost my father in body and my mother in mind at the same time. At times, it was so dark I couldn't see and could scarcely breathe. I lift up mine eyes and thank Jah moon and starshine for lighting the path.

And to my dad: I have kept the faith.

NOTES

PROLOGUE

1. The Nation's Report Card, 2024, https://www.nationsreportcard.gov/reading/nation/achievement/?grade=12.

2. Pooja Salhotra, "A.I.-Driven Education: Founded in Texas and Coming to a School Near You," *New York Times*, July 27, 2025, https://www.nytimes.com/2025/07/27/us/politics/ai-alpha-school-austin-texas.html.

3. Khanmigo, https://www.khanmigo.ai/learners, accessed July 18, 2025.

4. Chester Holland, Akisha Osei Sarfo, Mary M. Oewel, Casey A. Crews, Kyuna Sims, Brian Garcia, and Ray Hart, *Urban District Superintendents: Characteristics, Tenure, and Salaries*, Council of the Great City Schools, March 2025.

CHAPTER 1: HALLUCINATION AND SEGREGATION

1. James Baldwin, "A Talk to Teachers," October 16, 1963, https://www.zinnedproject.org/materials/baldwin-talk-to-teachers.

2. Ileana Najarro, "Here's What Happened with AP African American Studies in Georgia," *Education Week*, August 19, 2024, https://www.edweek.org/teaching-learning/heres-what-happened-with-ap-african-american-studies-in-georgia/2024/08.

3. Executive Office of the President, Executive Order 14110, "Safe, Secure, and Trustworthy Development and Use of Artificial Intelligence," November 1, 2023, https://www.federalregister.gov/documents/2023/11/01/2023–24283/safe-secure-and-trustworthy-development-and-use-of-artificial-intelligence.

4. W. E. B. Du Bois, "Strivings of the Negro People," *Atlantic Monthly*, August 1897, https://www.theatlantic.com/magazine/archive/1897/08/strivings-of-the-negro-people/305446/.

5. "Florida's State Academic Standards – Social Studies, 2023" in Rule 6A-1.09401, Student Performance Standards, Florida Department of Education, p. 6. https://www.fldoe.org/core/fileparse.php/20653/urlt/6-4.pdf.

6. Deborah Yao, "25 Years Ago Today: How Deep Blue vs. Kasparov Changed AI Forever," *AI Business*, May 11, 2022, https://aibusiness.com/ml/25-years-ago-today-how-deep-blue-vs-kasparov-changed-ai-forever.

7. Andrew R. Chow, "How ChatGPT Managed to Grow Faster Than TikTok or Instagram," *Time*, February 8, 2023, https://time.com/6253615/chatgpt-fastest-growing/.

8. Safiya Umoja Noble, *Algorithms of Oppression: How Search Engines Reinforce Racism* (NYU Press, 2018); Joy Buolamwini, *Unmasking AI: My Mission to Protect What Is Human in a World of Machines* (Random House, 2023).

9. Langston Hughes, "I, Too," in *The Weary Blues* (Knopf, 1926).

10. Derrick Bell, *Silent Covenants: Brown v. Board of Education and the Unfulfilled Hopes for Racial Reform* (Oxford University Press, 2005); Matthew Delmont, *Why Busing Failed: Race, Media, and the National Resistance to School Desegregation* (University of California Press, 2016); Kevin M. Kruse, *White Flight: Atlanta and the Making of Modern Conservativism* (Princeton University Press, 2005).

11. Nikole Hannah-Jones, "Our democracy's founding ideals were false when they were written. Black Americans have fought to make them true," *New York Times Magazine*, August 14, 2019.

12. David Margolick, *Elizabeth and Hazel: Two Women of Little Rock* (Yale University Press, 2011).

13. Eliza Shapiro, "Segregation Has Been the Story of New York City's Schools for 50 Years," *New York Times*, March 26, 2019; Emily Piper-Vallillo, "How School Segregation Survived Boston's Busing," WBUR, June 19, 2024; Sydney Grundland Lanuza, "A Legacy of Separation: Connecting San Francisco's History of Redlining to the Educational Disparities in SFUSD," *Berkeley Political Review*, April 22, 2024, https://bpr.studentorg.berkeley.edu/2024/04/22/a-legacy-of-separation-connecting-san-franciscos-history-of-redlining-to-the-educational-disparities-in-sfusd/.

14. Charles E. Cobb Jr. and Robert P. Moses, *Radical Equations: Civil Rights from Mississippi to the Algebra Project* (Beacon Press, 2002).

15. APS Insights, https://apsinsights.org/data-explorer/, accessed October 23, 2024.

16. Tanya Milberg, "Why AI Literacy Is Now a Core Competency in Education," World Economic Forum, May 22, 2025, https://www.weforum.org/stories/2025/05/why-ai-literacy-is-now-a-core-competency-in-education/.

17. Nadine Diaz-Infante, Michael Lazar, Samvitha Ram, and Austin Ray, "Demand for Online Education Is Growing. Are Providers Ready?" McKinsey & Company, July 20, 2022.

18. Natalie Schwartz, "Two-Thirds of Colleges Are Adding Online Programs, Survey Finds," *Higher Ed Dive*, August 15, 2023.

19. Michael Levenson, "Lori Loughlin Begins 2-Month Sentence for Role in Admissions Scandal," *New York Times*, October 30, 2020.

20. Ethan Bakuli and Micah Walker, "Detroit Students Show Slight Gains on Michigan's Standardized Test," *Chalkbeat Detroit*, September 1, 2023, https://www.chalkbeat.org/detroit/2023/9/1/23855803/detroit-public-schools-charter-mstep-test-scores-2023/.

21. Families in Action for Quality Education, *The Unspoken Pandemic: Oakland's Education Crisis*, 2023.

22. The Nation's Report Card, *2024 Mathematics Trial Urban District Snapshot Report*, 2024.

23. The Nation's Report Card, *2024 Mathematics State Snapshot Report*, 2024.

24. Sean F. Reardon, Ericka S. Weathers, Erin M. Fahle, Heewon Jang, and Demetra Kalogrides, "Is Separate Still Unequal? New Evidence on School Segregation," Working Paper No. 19–06, Stanford Center for Education Policy Analysis,

August 2022, https://cepa.stanford.edu/content/separate-still-unequal-new-evidence-school-segregation-and-racial-academic-achievement-gaps.

25. David Card, Alexandre Mas, and Jesse Rothstein, "Tipping and the Dynamics of Segregation," *Quarterly Journal of Economics* 123, no. 1 (February 2008): 177–218, https://doi.org/10.1162/qjec.2008.123.1.177.

26. US Census Bureau, "Quick Facts Atlanta City, Georgia," 2025, https://www.census.gov/quickfacts/fact/table/atlantacitygeorgia/PST045224.

27. Bruce C. Mitchell, Jad Edlebi, Helen C. S. Meier, Jason Richardson, Joseph Dean, and Liang Chen, *Displaced by Design: Fifty Years of Gentrification and Black Cultural Displacement in US Cities*, National Community Reinvestment Coalition, May 2025, https://ncrc.org/displaced-by-design/.

28. US Census Bureau, *Georgia: 2000 Summary Population and Housing Characteristics*, 2002, https://www2.census.gov/library/publications/2002/dec/phc-1-12.pdf.

29. Richard J. Herrnstein and Charles Murray, *The Bell Curve: Intelligence and Class Structure in American Life* (Free Press, 1996).

30. Ramin Skibba, "The Disturbing Resilience of Scientific Racism," *Smithsonian*, May 20, 2019, https://www.smithsonianmag.com/science-nature/disturbing-resilience-scientific-racism-180972243/.

CHAPTER 2: DEMOCRATIZING COMPUTING CAME FIRST

1. Google, *2024 Diversity Annual Report*, 2024, https://belonging.google/diversity-annual-report/2024/.

2. Facebook, *Annual Diversity Report*, July 2021, https://about.fb.com/wp-content/uploads/2021/07/Facebook-Annual-Diversity-Report-July-2021.pdf.

3. Maya King, "What's a 'Black Job'? Trump's Anti-Immigration Remarks Are Met with Derision," *New York Times*, June 28, 2024, https://www.nytimes.com/2024/06/28/us/politics/black-job-trump-immigration.html.

4. EdTrust, "Improving Students' Access to Advanced Coursework," December 7, 2023, https://edtrust.org/rti/improving-students-access-to-advanced-coursework/.

5. Steve Lohr, "'00 Computer Glitches Are Mostly a No-Show," *New York Times*, January 1, 2000, https://archive.nytimes.com/www.nytimes.com/library/tech/00/01/biztech/articles/01year.html.

6. Robert P. Moses and Charles E. Cobb Jr., *Radical Equations: Civil Rights from Mississippi to the Algebra Project* (Beacon Press, 2002).

7. Jonathan Kozol, *Savage Inequalities: Children in America's Schools* (Crown, 1991).

8. Carter G. Woodson, *The Mis-Education of the Negro* (Associated Publishers, 1933); Jarvis R. Givens, *Fugitive Pedagogy: Carter G. Woodson and the Art of Black Teaching* (Harvard University Press, 2021).

9. National Center for Education Statistics, "Fast Facts: SAT Score," 2024, https://nces.ed.gov/fastfacts/display.asp?id=171.

10. Nelson Mandela, *Long Walk to Freedom: The Autobiography of Nelson Mandela* (Back Bay Books, 1995).

11. Barack Obama, State of the Union Address, January 2016, https://obama whitehouse.archives.gov/the-press-office/2016/01/12/remarks-president-barack-obama-%E2%80%93-prepared-delivery-state-union-address.

12. Executive Office of the President, President's Council of Advisors on Science and Technology, *Report to the President, Engage to Excel: Producing One Million*

Additional College Graduates with Degrees in Science, Technology, Engineering, and Mathematics, 2012.

13. STEM Education Act of 2015, Public Law 114–59, 114th Congress, October 7, 2015, https://www.congress.gov/114/plaws/publ59/PLAW-114publ59.pdf.

14. Georgia Senate Bill 108, https://legiscan.com/GA/bill/SB108/2019

15. The Nation's Report Card, *2019 Mathematics State Snapshot Report*, https://nces.ed.gov/nationsreportcard/subject/publications/stt2019/pdf/2020013NP8.pdf, accessed July 10, 2024.

16. Jane Margolis, Jennifer Jellison Holme, Joanna Goode, Kimberly Nao, and Rachel Estrella, *Stuck in the Shallow End: Education, Race, and Computing* (MIT Press, 2008).

17. College Board Newsroom, "New Data: AP Computer Science Principles Course Bringing More Diverse Set of Students into Computer Science Pipeline," December 9, 2020, https://newsroom.collegeboard.org/new-data-ap-csp-course-bringing-more-diverse-set-students-computer-science-pipeline.

CHAPTER 3: THE EXCEPTION AND THE PATTERN

1. Angela Duckworth, *Grit: The Power of Passion and Perseverance* (Scribner, 2016).

2. James Baldwin, "A Letter to My Nephew," *Progressive Magazine*, December 1, 1962, https://progressive.org/magazine/letter-nephew/.

3. Langston Hughes, "I, Too," in *The Collected Works of Langston Hughes* (University of Missouri Press/BkMk Press, 2000).

4. Eliza Shapiro and K. K. Rebecca Lai, "How New York's Elite Public Schools Lost Their Black and Hispanic Students," *New York Times*, June 3, 2019, https://www.nytimes.com/interactive/2019/06/03/nyregion/nyc-public-schools-black-hispanic-students.html.

5. Ali Vitali, Kasie Hunt, and Frank Thorp V, "Trump Referred to Haiti and African Nations as 'Shithole' Countries," NBC News, January 11, 2018, https://www.nbcnews.com/politics/white-house/trump-referred-haiti-african-countries-shithole-nations-n836946.

6. Lisa W. Foderaro, "Angered by Attack, Trump Urges Return of the Death Penalty," *New York Times*, May 1, 1989, https://www.nytimes.com/1989/05/01/nyregion/angered-by-attack-trump-urges-return-of-the-death-penalty.html.

7. The White House, Executive Order 14173, "Ending Illegal Discrimination and Restoring Merit-Based Opportunity," January 21, 2025, https://www.whitehouse.gov/presidential-actions/2025/01/ending-illegal-discrimination-and-restoring-merit-based-opportunity/.

8. The White House, Executive Order 11246, "Equal Employment Opportunity," September 28, 1965, https://www.eeoc.gov/history/executive-order-no-11246.

9. Supreme Court of the United States, *Fisher v. University of Texas*, June 23, 2016, https://www.oyez.org/cases/2015/14-981.

10. Krista Cox, "Supreme Court Hears Oral Arguments in Fisher v. University of Texas (Fisher II)," *ARL Views*, December 9, 2015, https://www.arl.org/blog/supreme-court-hears-oral-arguments-in-fisher-v-university-of-texas-fisher-ii/.

11. James Henry Hammond, "Mud Sill Speech," March 4, 1858, https://teachingamericanhistory.org/document/mud-sill-speech/.

12. Sunny Nagpaul, "The South Bronx Is a Cross Section of Everything Wrong with America, but Has a Community Hellbent on Changing That: 'You Need to Have an Outlet So It Doesn't Destroy Your Life,'" *Fortune*, June 29, 2024, https://fortune.com/2024/06/29/south-bronx-poorest-congressional-district-america-poverty-pollution-economy/.

13. Quentin Hardy, "Jesse Jackson Confronts Silicon Valley," *New York Times*, March 19, 2014, https://archive.nytimes.com/bits.blogs.nytimes.com/2014/03/19/jesse-jackson-confronts-silicon-valley/.

14. UrbanGeekz Staff, "Jesse Jackson's Open Letter to the Tech Industry," *UrbanGeekz*, August 13, 2015, https://urbangeekz.com/2015/08/jesse-jacksons-open-letter-to-the-tech-industry/.

15. Malcolm X, "The Ballot or the Bullet," speech, King Solomon Baptist Church, Detroit, MI, April 12, 1964, https://bpb-us-e2.wpmucdn.com/sites.middlebury.edu/dist/0/2081/files/2013/02/Malcolm-X-The-Ballot-or-the-Bullet.pdf.

16. Sander Vanocur, "MLK Talks 'New Phase' of Civil Rights Struggle, 11 Months Before Assassination," May 8, 1967, posted Apr 4, 2018, by NBC News, YouTube, https://youtu.be/2xsbt3a7K-8?si=bKDm9fiaO_V2MBIS.

17. Google, *2024 Diversity Annual Report*, 2024, p. 78, https://belonging.google/diversity-annual-report/2024/.

18. "Best Computer Science Schools," *U.S. News & World Report*, 2025, https://www.usnews.com/best-graduate-schools/top-science-schools/computer-science-rankings.

19. MIT Schwarzman College of Computing, "MIT Reshapes Itself to Shape the Future," History, https://computing.mit.edu/about/history/, accessed December 8, 2025.

20. Integrated Postsecondary Education Data System (IPEDS), "Use the Data," National Center for Education Statistics, US Department of Education, https://nces.ed.gov/ipeds/use-the-data, retrieved March 19, 2025.

21. Katie Siek and Jasmine Batten, *Taulbee Survey 2024: Annual Report*, Computing Research Association, 2024.

22. Integrated Postsecondary Education Data System (IPEDS), "Use the Data."

23. Nate Silver, "The Most Diverse Cities Are Often the Most Segregated," *FiveThirtyEight*, May 1, 2015, https://fivethirtyeight.com/features/the-most-diverse-cities-are-often-the-most-segregated/.

24. NYU Furman Center, "Brooklyn Neighborhood Indicators," last updated May 8, 2025, https://furmancenter.org/neighborhoods/view/brooklyn#demographics.

25. NYC Chancellor's Office, *2020–21 School Year Financial Transparency Report: PS 321 William Penn*, https://data.nysed.gov/expenditures.php?year=2021&instid=800000045158.

26. Kahlil Gibran, *The Prophet* (Knopf, 1923).

CHAPTER 4: DIVERSITY DEFANGED

1. Derrick Bell, *Faces at the Bottom of the Well: The Permanence of Racism* (Basic Books, 1992).

2. Constitution of the United States, Article I, Section 2, https://constitution.congress.gov/constitution/article-1/.

3. United States Supreme Court, *Dred Scott v. Sandford* (1857), https://www.archives.gov/milestone-documents/dred-scott-v-sandford.

4. Constitution of the United States, Thirteenth Amendment, Section 1, https://constitution.congress.gov/constitution/amendment-13/.

5. Plessy v. Ferguson (1896), https://www.archives.gov/milestone-documents/plessy-v-ferguson.

6. Brown v. Board of Education (1954), https://www.archives.gov/milestone-documents/brown-v-board-of-education.

7. San Antonio Independent School District v. Rodriguez (1973), https://www.oyez.org/cases/1972/71-1332.

8. Voting Rights Act (1965), https://www.archives.gov/milestone-documents/voting-rights-act.

9. Shelby County v. Holder (2013), https://www.oyez.org/cases/2012/12-96.

10. W. E. B. Du Bois, *The Souls of Black Folk: Essays and Sketches* (A. C. McClurg & Co., 1903).

11. James Baldwin, *No Name in the Street* (Dial Press, 1972).

12. James Allen, John Lewis, Leon F. Litwack, and Hilton Als, *Without Sanctuary: Lynching Photography in America* (Twin Palms, 1999).

13. Domenico Montanaro, "A Wild Year in Politics, by the Numbers," NPR, December 27, 2024, https://www.npr.org/2024/12/27/nx-s1-5222570/2024-politics-recap.

14. Bell, *Faces at the Bottom of the Well.*

15. Charlie Rose Conversations, "Conversation with Toni Morrison," May 7, 1993, https://charlierose.com/videos/31212.

16. Isabel Wilkerson, *Caste: The Origins of Our Discontents* (Random House, 2020), 70.

17. Google, *2024 Diversity Annual Report*, 2024, https://belonging.google/diversity-annual-report/2024/.

18. Klaus Schwab, "The Fourth Industrial Revolution: What It Means, How to Respond," World Economic Forum, January 14, 2016, https://www.weforum.org/stories/2016/01/the-fourth-industrial-revolution-what-it-means-and-how-to-respond/.

19. Integrated Postsecondary Education Data System (IPEDS), "Use the Data," National Center for Education Statistics, US Department of Education, https://nces.ed.gov/ipeds/use-the-data, retrieved March 19, 2025.

20. Ember Smith and Richard V. Reeves, "SAT Math Scores Mirror and Maintain Racial Inequity," Brookings Institution, December 1, 2020, https://www.brookings.edu/articles/sat-math-scores-mirror-and-maintain-racial-inequity/.

21. Aatish Bhatia, Irineo Cabreros, Asmaa Elkeurti, and Ethan Singer, "Trump Has Cut Science Funding to Its Lowest Level in Decades," *New York Times*, May 22, 2025, https://www.nytimes.com/interactive/2025/05/22/upshot/nsf-grants-trump-cuts.html.

22. Chinese Exclusion Act (1882), National Archives, https://www.archives.gov/milestone-documents/chinese-exclusion-act.

23. William H. Frey, "New 2020 Census Data Shows an Aging America and Wide Racial Gaps Between Generations," Brookings, August 1, 2023, https://www.brookings.edu/articles/new-2020-census-data-shows-an-aging-america-and-wide-racial-gaps-between-generations/.

24. US Census Bureau, "New Estimates Highlight Differences in Growth Between the U.S. Hispanic and Non-Hispanic Populations," press release no. CB24-109, June 27, 2024, https://www.census.gov/newsroom/press-releases/2024/population-estimates-characteristics.html.

25. Melissa Sanchez, Jodi S. Cohen, T. Christian Miller, Sebastian Rotella, and Mariam Elba, "'I Lost Everything': Venezuelans Were Rounded Up in a Dramatic Midnight Raid but Never Charged with a Crime," ProPublica, November 13, 2025, https://www.propublica.org/article/chicago-venezuela-immigration-ice-fbi-raids-no-criminal-charges.

26. Simon Romero, Prior Beharry, and Federico Rios, "A Mystery in Trinidad as Bodies Wash Ashore After U.S. Strikes," *New York Times*, October 23, 2025, https://www.nytimes.com/2025/10/23/world/americas/trinidad-bodies-us-strike-venezuela.html.

CHAPTER 5: MANDELA AND ALTMAN

1. Donovan X. Ramsey, *When Crack Was King: A People's History of a Misunderstood Era* (One World, 2023).

2. Trinh Q. Truong, Debu Gandhi, and Jill Rosenthal, "Tackling the Opioid Crisis Requires a Whole-of-Government, Society-Wide Approach," Center for American Progress, September 20, 2023, https://www.americanprogress.org/article/tackling-the-opioid-crisis-requires-a-whole-of-government-society-wide-approach/.

3. *The Central Park Five*, PBS, April 15, 2013, https://www.pbs.org/video/central-park-five-media-response/.

4. "Bring Back the Death Penalty," Donald J. Trump's full-page ad, published in the *New York Times*, *Daily News*, *New York Post*, and *New York Newsday* May 1, 1989, during the media trial of the Central Park Five, https://www.reddit.com/r/PropagandaPosters/comments/5rltoh/bring_back_the_death_penalty_donald_j_trumps/, accessed July 22, 2025.

5. Jennifer Steinhauer, "Giuliani's Comments Part of a Complicated History on Race," *New York Times*, February 20, 2015, https://archive.nytimes.com/www.nytimes.com/politics/first-draft/2015/02/20/giulianis-comments-part-of-a-complicated-history-on-race/.

6. Nelson Mandela, "On Freedom," Pretoria Supreme Court, April 20, 1964, https://www.un.org/en/events/mandeladay/court_statement_1964.shtml.

7. Malcolm X, "Speech at the Founding Rally of the Organization of Afro-American Unity" (1964), *BlackPast*, October 16, 2007, https://www.blackpast.org/african-american-history/1964-malcolm-x-s-speech-founding-rally-organization-afro-american-unity/.

8. Adam Satariano and Paul Mozur, "The Global A.I. Divide," *New York Times*, June 21, 2025, https://www.nytimes.com/interactive/2025/06/23/technology/ai-computing-global-divide.html.

9. Karen Hao, *Empire of AI: Dreams and Nightmares in Sam Altman's OpenAI* (Penguin Press, 2025).

10. Hao, *Empire of AI*, 26.

11. Dario Amodei, "A Statement from Dario Amodei on Anthropic's Commitment to American AI Leadership," Anthropic, October 21, 2025, https://www.anthropic.com/news/statement-dario-amodei-american-ai-leadership.

12. The White House, Executive Order 14319, "Preventing Woke AI in the Federal Government," July 23, 2025, https://www.whitehouse.gov/presidential-actions/2025/07/preventing-woke-ai-in-the-federal-government/.

13. The White House, "Winning the Race: America's AI Action Plan," July 2025, https://www.whitehouse.gov/wp-content/uploads/2025/07/Americas-AI-Action-Plan.pdf.

14. Anthropic, "Measuring Political Bias in Claude," November 13, 2025, https://www.anthropic.com/news/political-even-handedness.

15. Marc Andreessen, "The Techno-Optimist Manifesto," Andreessen Horowitz, October 16, 2023, https://a16z.com/the-techno-optimist-manifesto/.

16. Tanya Milberg, "The Future of Learning: How AI Is Revolutionizing Education 4.0," World Economic Forum, April 28, 2024, https://www.weforum.org/stories/2024/04/future-learning-ai-revolutionizing-education-4-0/.

17. Natasha Singer, "Microsoft Pledges $4 Billion Toward A.I. Education," *New York Times*, July 9, 2025, https://www.nytimes.com/2025/07/09/business/microsoft-ai-education.html; Alyson Klein, "Google Spends $25M on AI Training for Teachers," *Education Week*, September 18, 2024, https://www.edweek.org/technology/google-spends-25m-on-ai-training-for-teachers/2024/09.

18. Human-Centered Artificial Intelligence, "Annual AI+Education Summit 2025: Human-Centered AI for a Thriving Learning Ecosystem," Stanford University, February 25, 2025, https://hai.stanford.edu/events/human-centered-ai-for-a-thriving-learning-ecosystem.

19. Allan C. Golston, "Unleashing AI's Potential for Equitable Academic Outcomes and Economic Opportunity," Gates Foundation US Program, December 17, 2024, https://usprogram.gatesfoundation.org/news-and-insights/articles/unleashing-ais-potential-for-equitable-academic-outcomes-and-economic-opportunity.

20. AI for Education, https://www.aiforeducation.io/.

21. Microsoft, *2025 AI in Education: A Microsoft Special Report*, 2025, https://cdn-dynmedia-1.microsoft.com/is/content/microsoftcorp/microsoft/bade/documents/products-and-services/en-us/education/2025-Microsoft-AI-in-Education-Report.pdf.

22. aiEDU, https://www.aiedu.org/.

23. Code.org, https://code.org/en-US.

24. Khanmigo, https://www.khanmigo.ai/teachers.

25. Integrated Postsecondary Education Data System (IPEDS), "Use the Data," National Center for Education Statistics, US Department of Education, https://nces.ed.gov/ipeds/use-the-data, retrieved March 19, 2025.

26. Char Adams, "MIT's Black Student Enrollment Drops Significantly After Supreme Court Affirmative Action Ruling," NBC News, August 21, 2024, https://www.nbcnews.com/news/nbcblk/mits-black-student-enrollment-slides-affirmative-action-supreme-court-rcna167622.

27. Sofia Williams, "Preliminary Admissions Data for Class of 2028 Shows Drop in Black, Latino Enrollment," *Stanford Daily*, October 2, 2024, https://stanforddaily.com/2024/10/02/class-of-2028-admissions-data/.

28. Bruce C. Mitchell, Jad Edlebi, Helen C. S. Meier, Jason Richardson, Joseph Dean, and Liang Chen, *Displaced by Design: Fifty Years of Gentrification and Black Cultural Displacement in US Cities*, National Community Reinvestment Coalition, May 2025, https://ncrc.org/displaced-by-design/.

29. Martha Dalton, "Atlanta Superintendent Wants to Take a 'Back to Basics' Approach," *Atlanta Journal-Constitution*, March 17, 2025, https://www.ajc.com/education/atlanta-superintendent-wants-to-take-a-back-to-basics-approach/GRE2QGUEQRDP3PIVX45RS732E4/.

30. Georgia Insights, "Computer Science," Georgia Department of Education, https://georgiainsights.gadoe.org/dashboards/computer-science/, accessed July 25, 2025

31. "Feb. 3, 1964: New York City School Children Boycott School," Zinn Education Project, https://www.zinnedproject.org/news/tdih/nyc-school-children-boycott-school/, accessed July 20, 2025.

32. James H. Cone, *The Cross and the Lynching Tree* (Orbis Books, 2011).

CHAPTER 6: EDUCATION VERSUS TRAINING

1. Nikole Hannah-Jones, *The 1619 Project: A New Origin Story* (One World, 2021).

2. Jake Silverstein, "The 1619 Project and the Long Battle over U.S. History," *New York Times Magazine*, November 9, 2021, https://www.nytimes.com/2021/11/09/magazine/1619-project-us-history.html.

3. Mariana Olaizola Rosenblat, "Palantir Is Profiting from Trump's Ravenous Appetite for Deportations," NYU Stern Center for Business and Human Rights, April 18, 2025, https://bhr.stern.nyu.edu/quick-take/palantir-is-profiting-from-trumps-ravenous-appetite-for-deportations/.

4. The White House, Executive Order 13958, "Executive Order on Establishing the President's Advisory 1776 Commission," November 2, 2020, https://trumpwhitehouse.archives.gov/presidential-actions/executive-order-establishing-presidents-advisory-1776-commission/.

5. Molefi Kete Asante, *100 Greatest African Americans: A Biographical Encyclopedia* (Prometheus Books, 2003).

6. Hannah-Jones, *The 1619 Project.*

7. Georgia Insights, "Computer Science," Georgia Department of Education, https://georgiainsights.gadoe.org/dashboards/computer-science/, accessed July 25, 2025.

8. "Florida's State Academic Standards – Social Studies, 2023," in Rule 6A-1.09401, Student Performance Standards, Florida Department of Education, p. 6, https://www.fldoe.org/core/fileparse.php/20653/urlt/6-4.pdf.

9. Ron DeSantis, Second Inaugural Address, January 3, 2023, https://www.flgov.com/eog/news/press/2023/governor-desantis-delivers-inaugural-address-sets-priorities-second-term.

10. Jonathan Raymond, "Kennesaw State Deactivating Black Studies Program, Two Others; Georgia NAACP Issues Condemnation & Asks to Meet with School Leadership," 11 Alive, April 30, 2025, https://www.11alive.com/article/news/education/kennesaw-state-deactivating-black-studies-program-georgia-naacp-responds/85-f836afd4-a086-4e19-8bcd-41d111c4ba89.

11. Rex M. Nettleford, *Inward Stretch, Outward Reach: A Voice from the Caribbean* (Macmillan Caribbean, 1993).

12. Muzzammil Mohabir, "An Overview of Guyana's Sugar Industry: Perspectives on Contraction and Prospects," *Journal of Academic Research and Essays*, August 24, 2023, https://jare.org.gy/overview-guyana-sugar-industry/.

13. Gaiutra Bahadur, "Is Guyana's Oil a Blessing or a Curse?" *New York Times*, March 30, 2024, https://www.nytimes.com/2024/03/30/headway/is-guyanas-oil-a-blessing-or-a-curse.html.

14. Cheikh Anta Diop, *Civilization or Barbarism: An Authentic Anthropology* (Lawrence Hill Books, 1991).

15. James Baldwin, *The Fire Next Time* (Vintage, 1962), 98.

16. Alex Camardelle and Jarryd Bethea, *Building a Beloved Economy: A Baseline and Framework for Building Black Wealth in Atlanta*, Atlanta Wealth Building Initiative, November 2023.

17. Atlanta Public Schools, "Grow with Google Career Certificates," https://sites.google.com/apsk12.org/apsgrowlcertificates/home?authuser=0, accessed June 28, 2025.

18. Statista, "Number of Full-Time Alphabet Employees from 2007 to 2024," https://www.statista.com/statistics/273744/number-of-full-time-google-employees/, accessed June 28, 2025.

19. National Science Board, "Trends in Undergraduate and Graduate S&E Degree Awards," in *Higher Education in Science and Engineering*, February 22, 2022, https://ncses.nsf.gov/pubs/nsb20223/trends-in-undergraduate-and-graduate-s-e-degree-awards, accessed June 28, 2025.

20. Grow with Google, "Generative AI for Educators with Gemini," https://grow.google/ai-for-educators/, accessed June 28, 2025.

21. "College Admissions Scandal: Complete Coverage of a Brazen Cheating Scheme," *New York Times*, https://www.nytimes.com/news-event/college-admissions-scandal, accessed June 28, 2025.

22. Learn4Life (Metro Atlanta Regional Education Partnership), *State of Education in Metro Atlanta Annual Report 2024*, https://l4lmetroatlanta.org/wp-content/uploads/L4L-Annual-Report-2024.pdf.

23. Samantha Dorisca, "Google Settles $50M Lawsuit That Alleges Bias Toward over 4,000 Black Employees," *AfroTech*, May 16, 2025, https://afrotech.com/google-settles-lawsuit-bias-black-employees.

CHAPTER 7: MODERN FREEDOM OR DIGITAL SLAVERY

1. Franky Abbott, "Cotton Gin and the Expansion of Slavery," Digital Public Library of America, https://dp.la/primary-source-sets/cotton-gin-and-the-expansion-of-slavery, accessed July 8, 2025.

2. History.com Editors, "Ku Klux Klan," History.com, October 29, 2009, last updated November 4, 2025, https://www.history.com/articles/ku-klux-klan.

3. "Racial Disparities Persist in Many U.S. Jails," Issue Brief, Pew Charitable Trusts, May 16, 2023, https://www.pew.org/en/research-and-analysis/issue-briefs/2023/05/racial-disparities-persist-in-many-us-jails.

4. Douglas A. Blackmon, *Slavery by Another Name: The Re-Enslavement of Black Americans from the Civil War to World War II* (Doubleday, 2008).

5. "16 June 1976 - Soweto Uprising," South African History Online, May 21, 2013, https://sahistory.org.za/article/june-16-soweto-youth-uprising.

6. Robert B. Edgerton, *Mau Mau: An African Crucible* (Free Press, 1989).

7. Anthony P. Carnevale and Jeff Strohl, *Separate & Unequal: How Higher Education Reinforces the Intergenerational Reproduction of White Racial Privilege*,

Center for Education and the Workforce, 2013, https://cew.georgetown.edu/wp-content/uploads/SeparateUnequal.ES_.pdf.

8. James Baldwin, "A Talk to Teachers," October 16, 1963, https://www.zinnedproject.org/materials/baldwin-talk-to-teachers.

9. James Baldwin, *The Fire Next Time* (Vintage, 1962).

10. René Descartes, *Discourse on Method and the Meditations* (1637) (Penguin, 1968).

INDEX